MAKING CHARCOAL AND BIOCHAR

A COMPREHENSIVE GUIDE

Jack Allonby DAVID JONES

Making Charcoal and Biochar

A Comprehensive Guide

Rebecca Oaks

THE CROWOOD PRESS

First published in 2018 by
The Crowood Press Ltd
Ramsbury, Marlborough
Wiltshire SN8 2HR

enquiries@crowood.com

www.crowood.com

This impression 2021

British Library Cataloguing-in-Publication Data
A catalogue record for this book is available from the British
Library.

ISBN 978 1 78500 399 8

Disclaimer
The author and the publisher do not accept any responsibility
in any manner whatsoever for any error or omission, or any loss,
damage, injury, adverse outcome, or liability of any kind incurred
as a result of the use of any of the information contained in this
book, or reliance upon it.

Designed and typeset by Guy Croton Publishing Services,
West Malling, Kent

Printed and bound in India by Replika Press Pvt Ltd

Contents

Dedication   6
Acknowledgements   6
Foreword by Don Kelley   7

 1 INTRODUCTION   8
 2 HOW CHARCOAL IS MADE   20
 3 EARTH BURNS AND PIT KILNS   30
 4 METAL KILNS AND BRICK KILNS   52
 5 RETORTS   78
 6 BAGGING AND MARKETING CHARCOAL   100
 7 LEGISLATION AND REGULATION   118
 8 COOKING WITH CHARCOAL   128
 9 BIOCHAR AND CARBON SEQUESTRATION   142
10 A TO Z OF CHARCOAL PRODUCTS   154

Appendix: Comparative Table of Charcoal Production   168
Glossary   169
Bibliography   171
Useful Addresses   173
Index   175

Dedication

To Walter Lloyd, 1924–2018: charcoal burner, musician, and enthusiast.

This book is also dedicated to my parents: Priscilla Penfold, a woman who had the courage to live her life on her own terms; she was an inspiration and a guide; Raymond Leach, who championed all the best causes with enthusiasm and charisma.

Acknowledgements

I would like to thank all my colleagues, friends and family who have assisted me in many and varied ways in writing this book. In particular I would like to thank the following:

Don Kelley for support and advice, and for agreeing to contribute to the foreword.

Brian Crawley and my colleagues, particularly Duncan and Flo, who are involved in the Coppice Association North West for keeping alive the knowledge of Cumbrian earth-burn techniques and much more.

Special mention must be made of the European Charcoal Burners convention (EVK) delegation, Alan and Jo Waters (thanks for the loan of 'Soul of Fire'), Peter Jameson, Alan Sage, Jane Ponton, Lynne and Pete Etheridge and Jame Hookway (thanks for the lift).

Also to Peter Quelch for giving me access to his charcoal archive.

The Coppice Co-op, especially Sam and James, for my endless badgering for pictures and help.

My NCFed colleagues, especially Martin Hale, Helen Waterfield, Carolyn and Hugh for photos.

All the many charcoal burners whose brains I have picked, including Jim Bettle, David Hutchinson, Matt Williams, Craig Sams, Matt Lewis, Mark Lloyd, Nick Harris, Darren Hopkins, Dan Sumners, Ian Taylor, Michael Wallwork, Twiggy and Paula Keary, Tony Callaghan, Stuart Gunner, David Hunter.

Also Iain Loasby for giving me a thorough insight into the Exeter retort.

Ruth Thompson for the Kon-tiki demo, and being a whizz at coming up with ideas for the tricky letters in Chapter 10.

John Allonby (son of Jack) who gave me access to old photos.

David Jones for access to his photo library.

Gillian Slater for the temporary loan of a very beautiful diamond ring.

Lesley Edkins for lending me her barbecue.

Andrea Pentecost for a beautiful charcoal drawing.

My sister-in-law Sophie Masson for help contacting the French caves for permission to use their photos.

Zephyrine Barbarachild for a phenomenal job editing.

Helen Shacklady and Lynne Alexander for proof reading.

My partner Amanda for major editorial input and endless support; Caroline for her recipe; and last but not least, Roxy the dog for getting me out of the house at regular intervals.

Foreword

This is a comprehensive and outstanding study of a material of the utmost importance to humankind; a material that brought to us the means to smelt and refine metals, was the beginning of the chemical industry, and provided us with the means to filter our drinking water and the heat for cooking our food as well as for production of the beakers, pots and containers we used.

It is not a book that could have been undertaken lightly. Reading it, one can appreciate the dedication needed to set down the long story, which began four thousand centuries ago when a group of hominids living in the Zhoukondian cave in northern China developed means of making fire at will. Large quantities of charred bones and deep ash layers testified to their industry and the length of their stay.

They had in fact developed controllable means of light and heat, and taken thereby the first fundamental step toward power generation and modern industry.

The span of this story is breathtaking and it is well told. Neither set up as a technical book nor pandering to popular views, it is of interest and instruction to the general reader as well as the forest worker striding into the woods with an axe or saw. The writer has demonstrated deep knowledge and affection for the industry and is therefore to be thanked and congratulated on a job very well done.

Donald W. Kelley
September 2017

D.W. Kelley was managing director Shirley Aldred and Company Ltd from 1969 to 1983. He was Secretary and Chair of the National Association of Charcoal Manufacturers and chaired the British Charcoal Group in the 1990s.

Chapter 1
Introduction

Making charcoal is like engaging in alchemy, from the first time you fill a kiln with wood and set fire to it, tend it lovingly till the smoke runs 'blue', then wait patiently until the kiln is cold, and at last tentatively approach to prise open the lid to see what you have created. A pile of ash, or a charred stack of logs? Or best of all, a heap of almost perfectly formed charcoal holding the shape and form of the wood, but none of its weight or strength: a brittle, tinkling, glistening treasure trove of charcoal. Magic.

CARBON – THE BUILDING BLOCK OF LIFE

Engaging in charcoal making links us directly with our ancestors – it is the one human activity that can be traced back through the millennia with very few changes in its most fundamental form. In today's world the most pressing preoccupation is with carbon: too much CO2 in the atmosphere, too much reliance on mining fossil fuels for energy. Can charcoal offer any answers?

In this book we will examine this question. It is a practical book with many ideas for making your own charcoal, an aspirational book that reports on the passion of those who think charcoal can save the world – and perhaps above all it is a fact-filled book with all the things you never knew you wanted to know about charcoal. So whether you are a barbecue or a biochar fanatic (or both), we hope you will find all the information you need within these pages.

In this chapter we set the scene with a look at what charcoal actually is, take a quick romp through the historical uses, and end with a broad-brush look at how charcoal is made.

Newly made lumpwood charcoal.

Carbon

Charcoal is carbon, a form of amorphous (any shape) carbon produced by partially burning wood or, in fact, any organic matter. All living things are made of carbon: atomic number 6 on the periodic table of elements (6C) with six electrons, four of which are available to bond with other chemicals to create a vast diversity of life. Bonding with each other to create 'chains', they bond with hydrogen and/or oxygen to make the building blocks of

Carbon Atom

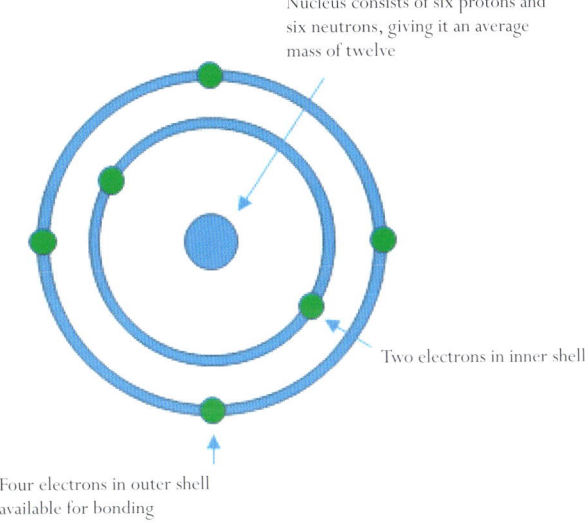

Nucleus consists of six protons and six neutrons, giving it an average mass of twelve

Two electrons in inner shell

Four electrons in outer shell available for bonding

The carbon atom.

Harder than Diamond

Lonsdaleite: Named in honour of the Irish crystallographer Kathleen Lonsdale. It forms when meteorites containing graphite impact with the Earth. The great heat and stress of the impact transforms the graphite into diamond, but retains graphite's hexagonal crystal lattice. Lonsdaleite may be transformed under pressure to become 58 per cent harder than diamond, a new world record.

Wurtzite Boron nitride: A name given to compounds with this same crystal structure, which can be compressed to form substances harder than diamond.

Buckminsterfullerene (C60): A spherical carbon molecule (Buckyball) or sometimes a tube. C60 solid is as soft as graphite, but when compressed to less than 70 per cent of its volume it transforms into a super-hard form of diamond.

Graphene: A form of graphite only a single cell thick. It is an amazing conductor of electricity, and one of the most recent additions to the 'harder-than-diamond club'. Russian-born physicists André Geim and Konstantin Novoselov were awarded a Nobel prize for its discovery in 2010.

Diamond ring.

all life on earth (and maybe elsewhere?). Ten million compounds in the universe involve carbon.

Carbon was one of the earliest elements identified by humans in ancient times. Three natural substances are pure carbon: the hardest is diamond – forged by extreme heat and pressure through geological processes to create one of the hardest, shiniest elements on earth (*see* box); graphite, which is put to good use in spreading and teaching the written word through the lead in pencils; and the softest form of carbon, charcoal.

Charcoal Structure

Charcoal differs structurally from the other two carbon forms. In graphite the molecules are joined to form two-dimensional structures which 'stack' together and shear apart as the pencil is pressed across the paper. Diamond molecules are orderly super-strong three-

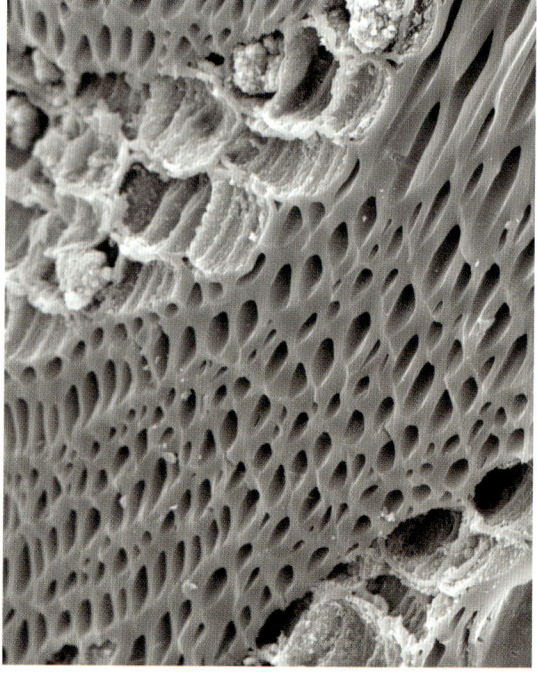

Magnified view of charcoal structure.
DARREN HOPKINS

dimensional bonds. In charcoal, molecules form a complex random matrix, which creates vast surface areas for moisture to cling to. One teaspoon of activated charcoal, if imagined as flattened out, has a surface area equivalent to that of a football pitch. It is hygroscopic, drawing moisture from the atmosphere and storing it within its cell structure. Any charcoal burner who has stored bags of charcoal over the winter in a damp climate will attest that it becomes heavy and damp and loses the ability to light readily with paper and a match.

CHARCOAL – THE BASICS

Pyrolysis

Turning any animal or vegetable matter back to pure carbon requires heat and combustion, which causes a chemical process called pyrolysis: the decomposition of complex substances into simpler ones by heating. The word 'pyrolysis' comes from the Greek *pyro* 'fire' and *lysis* 'separating'. The fire heats the substance and drives out the water and volatile compounds from the residual carbon and non-organic matter, leaving behind the pure carbon in the form of charcoal. Combustion requires oxygen, and in charcoal burning this element is carefully controlled. Only sufficient air is admitted in the heating process to raise the temperature through burning for the time it takes to complete combustion, then the oxygen is excluded once the distillation is complete. If heated too much, or for too long with too much oxygen present, the matter turns to ash.

Halting the process at just the right moment leaves a relatively pure residue of carbon. Don Kelly, one of Britain's great exponents of charcoal making, says 'Charcoal making is simply a function of time and temperature'. The enduring fascination of charcoal burning lies in learning the fine detail of how long and how hot to burn – an art that can encompass a lifetime of fine tuning to create the maximum quantity, best quality charcoal it is possible to produce in each burn.

Table of the Process of Pyrolysis

Kiln temperature	Distillates produced – variable dependant on the feedstock or timber used
20 to 110°C	Water vapour produced.
110 to 270°C	Residual water, some carbon monoxide, carbon dioxide, acetic acid and methanol given off. Heat absorbed (endothermic).
270 to 290°C	Exothermic decomposition of the wood begins (gives off heat). Heat is produced and as long as temperatures are maintained, breakdown continues spontaneously. Mixed gases and vapours produced; carbon monoxide, hydrogen and methane, some carbon dioxide and ethane as well as water, acetic acid, methanol, acetone and some tar.
290 to 400°C	Vapours include combustible gases carbon monoxide, hydrogen and methane and carbon dioxide gas and the condensable vapours; water, acetic acid, methanol, acetone. Also tars, which begin to dominate as the temperature rises which include, aldehydes, ketones, furfural, phenols.
400 to 500°C	The aim is to drive off the tars until the charcoal is a minimum of 75 per cent carbon, which requires the temperature to peak at 500°C.

The charcoal feedstock: This book focuses principally, but not exclusively, on hardwood timber as a feedstock for charcoal burning. Straw, sawdust and even sewage are also considered as potential sources of charcoal, both for briquettes and for biochar.

Dry distillation: Through the process of heating, the volatile compounds are driven from the wood, giving the process its formal name of 'dry distillation' or 'destructive distillation'.

Slow, Hot and Flash Pyrolysis

Slow Pyrolysis

Low temperatures and slow heating are known as slow pyrolysis, which creates a 'soft' or 'black' charcoal with a high density and high tar residues – so this type of material can fail to meet the carbon-content percentage required to call it charcoal (>50 per cent). Slow-burnt charcoal may retain a high volatile content, and will be heavier in weight and will behave differently to charcoal with a higher carbon content – it may be harder to light, but when up to temperature will burn more slowly and get hotter. Much imported charcoal falls into this category and is sold by weight, so the quantity in a bag may be much less compared to the same weight bag of charcoal produced at higher temperatures, and it will subsequently produce less heat overall. Some wood products sold as charcoal may be classified as 'torrified' wood, which is wood distilled at a temperature not exceeding 300°C.

Hot Pyrolysis

Charcoal produced at high temperatures (over 500°C) has a higher carbon content, of up to 98 per cent. An extreme form of this is known in Japan as 'white charcoal', or 'Binchotan', where it is highly prized for cooking but also has many industrial uses, including as a reducing agent for smelting. Charcoal produced at high temperatures has smaller pores but a greater pore volume or internal surface area, and even in its raw state has enhanced absorption, and is therefore more suitable for use as biochar (soil improver).

The very best quality charcoal with the highest pore volume is known as 'activated' charcoal, which has been super heat-treated with steam or chemicals to clean and open the pores to create a surface area in excess of 1,500sq m per gram, up to and over 3,000sq m per gram. This is the 'food grade' charcoal used in specialist filtration.

Flash Pyrolysis

In flash pyrolysis the feedstock is exposed to temperatures between 350°C and 550°C for less than two seconds, a process that produces more oils and tars, making it suitable for distilling biofuels; however, it creates less carbon volume.

The Properties of Charcoal

The main properties of charcoal are flammability, abrasion, absorption (where substances are drawn into the charcoal structure) and adsorption (where substances bond on the surface of the charcoal on the surface).

Flammability

The low moisture content and low residual volatiles create a substance that will ignite easily, and will burn cleanly and evenly without smoke and at a steady temperature, making it ideal for cooking and heating in a myriad of circumstances.

Absorption and Adsorption

The huge internal surface area and the molecular structure that encourages bonding to hydrocarbons (adsorption) makes charcoal ideal for filtration, for medicinal treatments that absorb toxins (absorption) and remove them from the body, and for holding nutrients in soils.

Abrasion

Because charcoal is non-soluble the particles remain intact even when immersed in water. This property has long been exploited for polishing (for example, copper printing rolls). More recently it has been adopted for cosmetic uses where finely ground charcoal is added to facial and body scrubs to provide a safe, eco-friendly form of abrasive. Most people making charcoal can attest that getting covered in charcoal dust gives you very soft skin, certainly after applying copious hot water and soap to remove it from sweaty or greasy skin.

CHARCOAL USES THROUGH THE AGES

Charcoal has been used extensively over thousands of years, as evidenced in prehistoric artwork and archaeological finds. Its uses include metallurgy, such as smelting copper and iron, in gunpowder production and wood distillates, and for cooking and medicines. In recent times charcoal uses include horticultural biochar and barbecues.

Early Uses of Charcoal

Human use of charcoal is recorded in prehistoric artwork and archaeology, dating back to the early use of fire, where the charred remains of wood on the cold hearth would have been collected to use for artwork to decorate and inform the prehistoric cave dweller. Examples of such cave art from 31,000 years ago have been found in France in the Ardèche at Vallon-Pont-D'Arc.

Chauvet Cave drawings. MINISTÈRE DE LA CULTURE ET DE LA COMMUNICATION

Reconstruction of an Iron Age Bloomery Furnace

As part of archaeology week at Hardcastle Crags in West Yorkshire, PhD student Hywel Lewis, in his research on charcoal burning, helped to create a reconstruction of an experimental bloomery furnace. This is the method whereby iron was smelted during the Iron Age right up to medieval times. This kind of furnace used charcoal made in an earth burn. The furnace shaft was constructed and prepared in advance from clay, sand and straw, then baked by a fire set inside to harden the clay. The reconstruction of the furnace at Hardcastle was based on material evidence from excavation.

Other PhD and Masters students, including Louis-Olivier Lortie, Yvette Marks and Benoit Proulx, from the University of Sheffield, and with the help of a number of willing volunteers, conducted the smelt. This involved preheating the bloomery, then filling it to the top with charcoal. Air was added by pumping bellows to bring the furnace up to temperature. At this point charges of crushed iron ore were added in layers alternating with charcoal at a set ratio of around two parts charcoal: one part ore.

This process was repeated over five hours, and used overall 66–88lb (30–40kg) of charcoal. After several hours of hard work, the door of the bloomery was broken open while the whole structure was still at high temperature, and the spongy 'bloom' or iron was extracted from the furnace. This raw, spongy metal was beaten with wooden mallets into a solid lump of iron.

Louis-Olivier Lortie constructing the bloomery furnace shaft. PENNINE PROSPECTS/STEVE MORGAN

RIGHT: *Benoit Proulx and a helper work the bellows.* PENNINE PROSPECTS/ STEVE MORGAN

FAR RIGHT: *Yvette A. Marks and Louis-Olivier Lortie hammer the spongy bloom.* PENNINE PROSPECTS/ STEVE MORGAN

Charcoal Use in Metallurgy

Circumstantial evidence links early metallurgy to charcoal production. Copper production indicates the use of charcoal to create the high temperatures required to smelt metal. The earliest dated find from Serbia is 7,500 years old. The earliest complete copper axe with a leather-bound yew handle was found on the 'ice man' Ötzi on the Austria/Italy border, dated at 5,300 years old. He may well have been a copper smelter himself, as traces of cyanide, a by-product of copper smelting, were found in his hair, an indication that he was probably involved in the trade. However, recent research suggests that the copper came from South Tuscany, far from where he was found, so either he was a traveller, or he traded his axe.

A replica of the ice axe carried by Ötzi.
ANDREAS FRANZKOWIAK

Charcoal Demand in the Iron Age

The addition of tin to copper to create bronze during the Bronze Age (2000BC–700BC) would have increased the need for charcoal. However, the biggest step-change in charcoal demand was in the Iron Age (700BC–43AD), because it would have been impossible to have smelted iron without considerable quantities of charcoal. Evidence of this production in Britain pre-dates the Romans with early 'bloomeries', a type of furnace where bellows were used to increase the burning temperature of the charcoal, and which have been found in the Sussex Weald, Kent, the Forest of Dean, Sherwood Forest, Northamptonshire and Furness (in what is now Cumbria).

Iron Production in the Roman Era

The Romans increased the production of iron in Britain and throughout their territories to service their great need for armour and weapons to expand and defend the empire. Oliver Rackham calculates that between 120 and 240AD in Sussex alone 550 tons of iron per year were produced. Using the formula that 7 tons of wood are required to make 1 ton of charcoal, and 12 tons of charcoal are required to make 1 ton of bar iron, a total estimated 50,000 tons of wood were required annually, obtained from an area of 25,000 acres (10,000ha) of woodland coppiced on a twelve-year rotation (giving a harvest of 24 tons per acre, or 60 tonnes per hectare). Interestingly, Rackham estimates that at that time the Sussex Weald had around 600,000 acres (243,000ha) of woodland, so only a small proportion (4 per cent) of the potential timber resource was used for iron production.

Blast Furnaces in the Production of Iron

Bloomery furnaces were the only way to obtain iron in Britain until the start of the sixteenth century, with the arrival of the blast furnace, which came into the Sussex Weald from France. Blast furnaces were more efficient than bloomery furnaces and the production of iron increased dramatically.

Demand for iron was high in the early eighteenth century, with eighty-five known sites in

A New Market for Charcoal

Hywel Lewis, discussing the need for charcoal in wool production, describes the moment when the arrival of coke caused the collapse of the market for charcoal supplying the iron industry in West Yorkshire. However, the industry did not disappear: a new market arose for charcoal, to heat the braziers for warming the wool and the combs used to straighten the fibres to create the fine worsted fabrics for which Yorkshire became famous. Lewis explains that:

A lock of Wensleydale longwool combed out for worsted spinning.

By 1800 the worsted trade had become centred on the three towns of Halifax, Bradford and Keighley, but production was still spread throughout their surrounding villages, hamlets and isolated farmsteads.

The introduction of mill working systems and the adoption of machinery for spinning and weaving heralded an explosion in the industry, from around seven million pounds of wool processed in 1800 to 80 million by 1850. Woolcombing, however, proved difficult to mechanise. This meant that the leviathan mills built on steam spinning and weaving, such as Salts Mill at Saltaire and Black Dyke Mills in Queensbury, were entirely

dependent on hand combers working two or three to a stove and combing thirty to fifty pounds of wool per week each.

From: Lewis, H. *Charcoal and wool: the role of woodland in the worsted revolution of West Yorkshire* (unpublished dissertation)

Supplying this trade would have kept many colliers at work, and demonstrates the adaptability of the woodland trades to meet changing demands over time.

Britain producing an estimated 24,000 tons of iron annually, using an eye-watering 432,000 tons of timber to make 62,000 tons of charcoal. Rackham calculated that this required 220,000 acres (89,000ha) of woodland, felled on a twelve-year cycle, representing a seventh of the total woodland in Britain.

The high level of demand for wood countrywide was ultimately unsustainable, given the rival pressures on land use for grazing and agriculture. Hence the development of coke or carbonized coal at the start of the eighteenth century was timely, as an alternative fuel for iron production. It took more than fifty years for coke to supersede charcoal for smelting, and charcoal as a fuel would never again be in such huge demand.

Some regions were slower than others to switch to coke, and in 1889 Backbarrow Iron-

works in Cumbria recorded using 58 tons of charcoal and 35 tons of iron ore to manufacture 28 tons of pig iron every week. 1926 was the final year when charcoal was used for smelting at Backbarrow.

Throughout Europe and the world there was a similar picture of iron production developing during the seventeenth and eighteenth centuries. Other metal-based industries such as steel and tin also made demands on charcoal. The effect on the landscape of the switch from wood to coal-based fuel must have been extreme.

Charcoal Production for Gunpowder

China is credited as being the birthplace of gunpowder, with written records from the eleventh century. Originally, two key ingredients were used medicinally: potassium nitrate (otherwise known as saltpetre) and sulphur. Someone must have had the bright idea to mix it with ground charcoal to create an explosive compound – and we can only hope they didn't blow themselves up in the process! These three ingredients in various ratios make up 'black powder', or gunpowder as we came to know it – a slow-burning, smoky product that was nevertheless more flammable than any other compound known to humankind at that time.

In China, people used gunpowder for fireworks before they thought of putting it to military use. By 1270, when Kublai Khan set out to conquer China, he had firearms and even grenades to further his expansionary aims. In other parts of the world the military soon adopted gunpowder: the first record of its use in battle by the British army was in 1346 at the Battle of Crecy. By the late sixteenth century, in Britain gunpowder mills were being bought and built by the British government and controlled by a royal warrant. During the English Civil War (1642–45) there was a major expansion of gunpowder mills, prompting the repeal of the royal warrant and the increase of private enterprise.

New gunpowder mills were sited where there was a ready supply of timber for charcoal production. While alder and willow were favourite species, alder buckthorn Frangula alnus was found to be the very best wood for gunpowder charcoal. This fairly modest shrub, found in wet woodlands, and widespread but not particularly common, was in high demand.

Like other charcoal products, its demand for gunpowder fell in the nineteenth century with the discovery of more efficient explosives, though there remains to this day a residual use for charcoal-based gunpowder in slow-burning fuses.

By-products of Charcoal Production

Pyroligneous acid (composed primarily of acetic acid, acetone, methanol and smaller amounts of more than 200 different organic compounds) – also known as 'wood vinegar' – is a dark, volatile product of wood distillation. It was discovered initially as a by-product of charcoal production, and developed both in conjunction with the gunpowder industry and in its own right. The importance of pyroligneous acid (as a source of

acetic acid) was known to the Egyptians, who used it for embalming the dead. When pyroligneous acid is distilled, 'wood spirit' or methanol is produced, which has been used to produce acetates of iron, lead, copper and calcium utilized in a wide range of industries, including dyeing and textile manufacture. Wood vinegar has been used as an insecticide and pest deterrent for centuries. The medieval name for this wood spirit was 'essence of smoke'.

Other by-products include 'wood tar' and 'wood oils'. The most famous of these is Stockholm tar from pine wood, which Pliny described in Roman times as a wood pitch for use in ship-building to caulk boats. Quantities of wood tar were found on an Etruscan ship that sank in 600BC and there is evidence of its use through the centuries, as with the famous sixteenth-century ship *Mary Rose*.

The volatile gas produced during wood distillation is the highly flammable hydrocarbon 'naphtha'. This gas is recycled to create the heat in retort burns.

Charcoal in Cooking

One of charcoal's amazing qualities is that it is a practically smokeless fuel. All the volatile oils are driven off during charcoal making, so charcoal gives an even heat and a clean burn which do not sting the cook's eyes with smoke. This superior fuel has been valued for domestic use since long before the advent of the modern barbecue. Add to this attribute its weight ratio of around 6:1 to 8:1, and it makes sense to convert wood to fuel close to production, and then transport the bulky but lightweight product to domestic hearths for cooking, a practice still common in many subsistence economies around the world.

The Cultural Significance of Charcoal

There is considerable mystery around charcoal burning, and various myths, superstitions and rituals are associated with its use and production. In twenty-first century Britain such mysteries have mostly disappeared. In Europe, a concerted effort has been made to keep the traditions going, even though traditional ways

Catering with charcoal at Le Château Fort du Fleckenstein.

of burning are no longer used for commercial production. In July 2017 the Europäischer Kolerverein (EKV: European Charcoal Burners, a group formed in 1997) assembled for their biennial meeting hosted by Les Charbonniers du Fleckenstein at Lembach, France, with over 300 people attending, all passionate about keeping the traditions of charcoal burning alive.

Interestingly, the Christian church adopted as the patron saint of charcoal burners Saint Alexander of Comana, a third-century bishop who was thought to have taken up charcoal burning to practise humility; his saint's day is celebrated on 11 August.

Chapter 2
How Charcoal Is Made

There are three main ways to make charcoal: in an earth kiln or pit kiln (known also as an earth burn); in a metal or brick kiln; or in a retort. Amongst new ideas in the charcoal-burning world is the Kon-tiki biochar kiln.

EARTH-COVERED KILNS, OR EARTH BURNS

The earliest charcoal burning is most likely to have been in open fires set in a pit, managed by 'flame curtain' pyrolysis (*see* Chapter 4), where the burning wood gases form a layer over the top of the lit timber, thus excluding oxygen, which prevents the developing charcoal from oxidizing and turning to ash. Burned in a conical pit, like those excavated in the Amazon basin, the fire was easy to control by constantly feeding it with wood, quenching the fire with water once the burn was finished.

Using earth to control the air supply may have developed in parallel to this, or indeed as an adaptation of the pit burn. The timber is packed into pits and covered with earth (*see* Chapter 3). These pit kilns would have had less

A traditional earth clamp in France.

The site of a charcoal burners' hut in the Lake District.

surface area to control in relation to the quantity of timber, but it would have proved harder to excavate the finished charcoal after the burn. The earth burn that stood above ground level almost certainly developed from this earlier type of pit burn, because even now the prepared ground, hearth or platform is still referred to as 'pitsteads' in some parts. Thus with very little variation, this technique has continued almost unchanged from prehistory to the present time.

The earth burn is a labour-intensive system and relatively inefficient, in that the ratio of charcoal produced to seasoned timber required is considerably less from an earth burn than from more modern methods such as kilns or retorts. In spite of this, the earth burn re-mains a way of adding value to timber with little more initial outlay than an axe, a shovel and a bucket. The earth burn still seems to hold a fascination: the romance of the woodland worker living night and day by the side of a living, glowing, smoking charcoal heap was captured in literature by Arthur Ransome in *Swallows and Amazons* (set in the English Lake District): this series of books has inspired young readers since the 1930s, and has been made into both a television series (1963) and two feature films (1974 and 2016).

Suddenly high in the darkness they saw a flicker of bright flame. There was another and then another, and then a pale blaze lighting a cloud of smoke. As they watched, the figure of

Jack Allonby outside a Lake District charcoal burners' hut. DAVID JONES

a man jumped into the middle of the smoke, a black, active figure, beating at the flames. The flames died down, and it was if a dark blind were drawn over the little window.

(*Swallows and Amazons*,
Arthur Ransome, 1930)

The charcoal burners' living conditions were perhaps not quite as romantic as they might have seemed from Ransome's description. They tended to have very basic accommodation in the woodlands, living in a charcoal burners' hut, a round structure with or without low walls and a tall pole conical structure covered in a thatch of bracken or turves. The remains of these huts are still found in some woodlands, as evidence of a lost way of life, though now only the stone slab hearths remain where the burners set a fire for heating and cooking.

Charcoal burning was mainly seasonal work, avoiding the worst of the winter weather, when the burners were likely to be busy felling timber. They may well have farmed smallholdings and had a settled life for part of the year. Memories of the lifestyle recorded by John Allonby, son of Jack Allonby who appeared in the *Swallows and Amazons* 1974 film, recalled good food and a fatted pig slaughtered for the charcoal-burning season. Despite their labour-intensive work, charcoal burners were often long-lived, and written records show they left fair financial legacies to their families, indicating that a good living could be made.

METAL AND BRICK CHARCOAL KILNS

In an attempt to make charcoal production more efficient, and to capture some of the waste products such as wood tar and pyroligneous acid, brick kilns were developed in the late eighteenth century – but by the nineteenth century the technology of brick kilns was largely superseded by metal retorts (*see* below). However, brick and concrete kilns are still found in Brazil and North America.

Coppicer Twiggy tending an 8ft (2.4m) metal ring kiln.

In recent times, the classic metal ring kiln is the most widespread way of making charcoal in Britain. Developed in the nineteenth century and promoted widely for use in Africa, it was not adopted by British charcoal burners until the late 1930s, when demand for charcoal recovered temporarily at the onset of World War II. It was undoubtedly a quicker way to make charcoal, and needed less constant attendance than an earth burn, cutting down on labour costs. Metal ring kilns provided a relatively small-scale option, whereby the kilns could be taken to the wood supply, gaining transportation efficiencies by converting wood to charcoal 'in situ', rather than carting heavy timber to static charcoal plants. Although the metal kiln still only offers a conversion rate of 1:6 at best, often 1:8 (similar to the earth burn), the method requires less labour.

Lyn Armstrong notes in her authoritative book *Wood Colliers and Charcoal Burning* that in 1961, of some 200 members of the National Association of Charcoal Manufacturers in Britain, about half were using portable metal kilns.

Don Kelley was the Chairman and Secretary of the National Association of Charcoal Manufacturers, which was active from 1926 to 1984. After retiring from that post, he continues to champion both small- and large-scale charcoal production to the present day. A rough estimate of the picture in 2017 suggests that there are still about 200 burners using metal kilns, but very few of those are producing significantly large quantities (an average of around 10 tonnes per year).

CHARCOAL RETORTS

The current charcoal-burning trend favours the retort. New portable versions are available, which combine the advantages of much greater efficiency with a conversion rate of 1:4, and considerably less pollution. The retort works by heating the wood in a separate chamber from the combustion, but the two chambers are linked so that when the wood reaches a sufficiently high temperature (270°C), the volatiles are released and piped back into the combustion chamber, creating the heat source to continue the burn. With combustion ending only when the volatiles are gone (at 450°C), the resultant charcoal is very good quality. Burning off this 'wood gas' reduces kiln emissions, releasing less smoke into the atmosphere.

Ger van Marion (known as GP), a renowned burner from the Netherlands and a member of the European Charcoal Group, refers to retorts as 'third generation' technology, consigning the earth burn and the metal kiln (first and second generation) to history, as inefficient and polluting. However, the retort design dates back to the 1800s and has been the main way of mass producing charcoal since the nineteenth century.

Charcoal burners in Britain are catching on to the efficiencies offered by retorts, and many of the larger producers and some smaller ones have switched to retorts despite the much higher initial capital outlay.

HEALTH WARNING

The various methods of making charcoal share an inherent danger. You are creating flammable gas, which can explode with great force. 99.9 per cent of the time you will be in control of it – but just occasionally accidents happen. The earth burn blows out a big hole that needs capping urgently, or the ring kiln gets so hot you can't get near the lid to settle it for flaring, or your retort just blows up – a very expensive and dangerous calamity, which can be avoided by resisting the temptation to take a peek. If you think you have shut down your clamp or kiln, do not be tempted to have a quick look, as the combination of heat, gas and oxygen is lethal. Even when you think your kiln is cool the admission of air can cause the charcoal to warm up again. At this point you must have water to hand to spray on and cool things down.

Wood gas flare under the lid of the kiln.

James Hookway with his charcoal retort in 2016.

THE KON-TIKI BIOCHAR KILN

The most recent innovation in the charcoal-burning world is the Kon-tiki biochar kiln, which was developed from the cone-shaped pits found in South America, in which it is surmised that the ancient cultures created charcoal used for improving the soil. The Kon-tiki kiln's main benefit is that you can burn brash or small-diameter waste wood – even wood chip – and keep filling the kiln with fuel until it is full of charcoal. The very hot burn (500°C+) leaves very few volatiles, producing an open, crumbly textured product that maximizes the porous nature of the charcoal, making it ideal for trapping nutrients in the soil (*see* Chapter 3 for a detailed description of the Kon-tiki kiln.)

Note: *See* the Appendix for a table comparing all the various methods of charcoal burning mentioned in this book. The table includes the initial cost, the timber requirements, and the yields of the various kilns discussed.

TIMBER FOR CHARCOAL MAKING

Any material, animal or vegetable, can be used to make charcoal. However, the focus of this book is primarily about making charcoal from timber, whether brushwood for biochar or massive chunks of wood in a metal ring kiln: the main ingredient comes from a tree. This also includes charcoal feedstock from a variety of vegetable sources such as coconut or coffee bean husk, or sawdust and wood chippings.

The Composition of Wood

Wood is composed of cellulose, hemicellulose and lignin, and volatile compounds including water, tannins and resins. Cellulose has a crystalline structure and forms long, string-like structures called microfibrils, which give wood its tensile strength and ability to bend. The microfibrils are encrusted with hemicellulose and lignin to make up the cell walls within the

A Kon-tiki kiln.

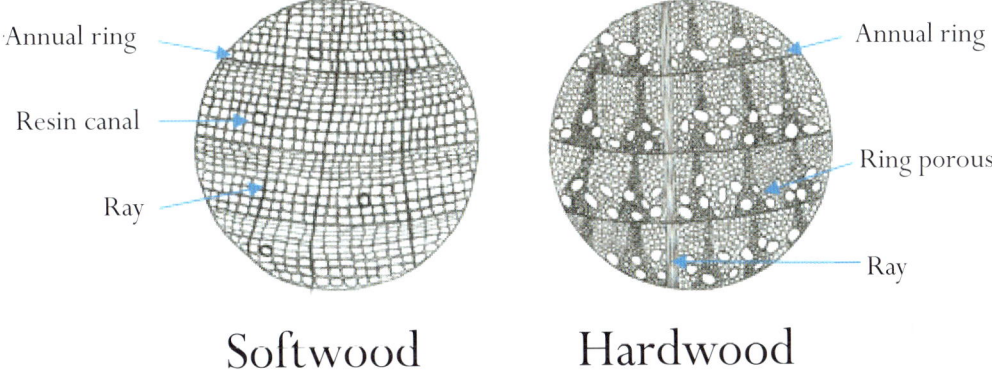

Softwood   Hardwood

Diagram of softwood and hardwood.

wood. Hemicellulose has a random structure and no inherent strength, while lignin is the strongest of all and fills the gaps between the cells, lending strength and 'stiffness' to the cell wall. Lignin contains aromatic phenols, which give character to the charcoal when their traces are still present.

Hardwood or Softwood

Hardwood and softwood both make charcoal perfectly well. Hardwoods are usually denser than softwoods on account of the timber's cell structure, which in turn affects the structure of the carbon produced. Softwoods have a simple cell structure, 90–95 per cent of which are tube-like tracheids that transport water and nutrients up the tree. These structures can break up under the stress of pyrolysis, creating a smaller particle size. Hardwoods have a more complex cell structure, with five or more types of cell, with greater variation throughout the growing period, giving a characteristic annual ring with early wood and late wood making up the varied grain patterns. This strong structure withstands the processes of pyrolysis and can remain virtually intact whilst being perfectly converted to charcoal.

The Moisture Content of Wood

Freshly felled, green wood can contain as much as 50 per cent water by weight. The amount of water in wood is expressed as moisture content, measured with a moisture meter. Air-dried over summer the moisture content will drop, especially if already prepared into suitable lengths for charcoal making. Air-dried timber, even if stored where the air can circulate and it is protected from rain, will be unlikely to have a moisture content of under 18 per cent. A moisture content considerably lower than 18 per cent is preferable for the most efficient conversion to charcoal, which can be achieved by some form of pre-drying, as seen in retort technology.

Shattered Charcoal

When there is too much moisture in the wood, or if the individual pieces of wood are too large when loaded into the charcoal kiln, a shattering effect can occur as the steam attempts to burst out of the wood. Low moisture content and a slower rise to temperature may prevent shattering. If your logs are large, cut them shorter (but don't necessarily split them) so the moisture can escape down the length of the grain.

Woodland Management

Charcoal burning in Britain goes hand in hand with sustainable woodland management. The hardwoods required for charcoal production are found in our woodlands, and these woods need managing. Many habitats for woodland flora and fauna are supported by well-managed woods, particularly the traditional form of woodland management known as coppicing. (*See* Oaks and Mills 2010.)

Coppicing

Coppicing is a form of woodland management which involves cutting down hardwood trees (not conifers) to almost ground-level, allowing the new shoots produced from the roots or the tree stump to develop into a multi-stemmed tree, which can be repeatedly cut and cropped without losing its vigour.

The drive to return more woodlands to coppice management has come from the conservation sector, which recognized that many woodland species are uniquely adapted to living in coppice woods. These have subsequently been under threat since the 1960s when commercial forestry (mainly conifer plantations) – regarded as a more efficient form of timber production – superseded the widespread coppicing system. Previously coppiced woods were either converted to conifer plantations or left to become derelict through lack of active management.

However, the forestry sector has recently started to take a second look at coppicing, realizing that the softwood plantations of the 1960s and 1970s often failed to provide the quality of timber needed – timber was typically produced abroad in greater quantities and more cheaply. However, commercial forestry practice in Britain has left a legacy of many thousands of hectares of unmanaged woodlands. The rebirth of demand for wood fuel encourages woodland owners to utilize this valuable resource – and whether by thinning or coppicing, the resultant timber is perfect for charcoal making.

Sources of Timber for Charcoal Burning

For small-scale operations you may be able to obtain some timber from a local arboriculture company (tree surgeon), or contact your local

Sycamore coppice, fifteen years old and ready for re-coppicing.

Wood stacks drying.

conservation group, doing woodland management, as there is often an opportunity to take timber for little or low cost, especially if you volunteer your time. With more and more people burning wood for heat, finding a free supply is difficult.

Felling your own timber is the next option, although this requires a certain financial outlay if you need to get the correct equipment, protective clothing and training for chainsaw use. But you can have the pleasure of seeing the process through from standing tree to bagged charcoal. However, felling efficiently and economically will take time and practice to do well, and unless you discount some of your own time, then this is not necessarily the best option financially.

Buying timber, either delivered or at the roadside or rideside (if you have an off-road vehicle suitable to collect it), will be the most likely way to proceed if you are doing more than just the occasional burn, and costs will vary depending on your location and the distances involved in transportation. The demand for timber is quite high at present, especially good hardwood suitable for domestic fires, and prices do reflect this, with £50 per tonne at the roadside not unusual. This is for green timber in long lengths, of course, and you do still have to process it (cut, split and dry it) before making charcoal.

The cost of your raw material will be the factor that determines profit or loss on a charcoal business, so thorough market research of prices and availability should be done locally before you begin.

Chapter 3
Earth Burns and Pit Kilns

EARTH BURNS

The noise of the chopping was now close at hand. A keen smell of smouldering wood tickled their nostrils. Suddenly they came out of the trees again on the open hillside. There were still plenty of larger trees, but the smaller ones and the undergrowth had been cut away. There were long piles of branches cut all of a length and neatly stacked, ready for the fire. There was one pile that made a complete circle with a hole in the middle of it. Forty or fifty yards away there was a great mound of earth with little jets of blue woodsmoke spurting from it. A man with a spade was patting the mound and putting a spadeful of earth wherever the smoke showed. Sometimes he climbed on the mound itself to smother a jet of smoke near the top of it. As soon as he closed one hole another jet of smoke would show itself somewhere else.

This vivid description of an earth burn from Arthur Ransome's *Swallows and Amazons* must surely have been written from a first-hand encounter with Lake District charcoal burners. He moved to the Lakes in 1924, and wrote this passage in the book, which was published in 1930. His arrival just overlapped with the final decline of the charcoal industry in the Lakes, as Backbarrow ironworks finally ceased charcoal production in 1926. Ransome reported that all charcoal burners kept an adder close by for luck – and we can assume that this was not necessarily artistic licence. In Cumbria it is often noted that somewhere very near a 'pitstead' you will almost always find a crab apple, a tree associated with folklore, adding to the mystery of charcoal-burning culture.

Fortunately, in the early 1970s traditional earth burning in Cumbria was recorded by two separate film-makers: Michael Dow in 1972 and Sam Hanna in 1973. They both filmed Cumbrian charcoal burners Jack and Bill Allonby building and firing a three-tiered charcoal stack about 60ft (18m) in diameter, recording the process in surprising detail. Jack was then asked to set up another burn for the 1974 film, *Swallows and Amazons*. This, though, was more a 'demonstration burn', with some artistic licence. In 1984, Arthur Barker resurrected earth burns at Brantwood on Coniston Water, enlisting Jack Allonby's help and advice, and later a burn at Ickenthwaite with Alan Waters from the Weald and Downland Museum. The Coppice Association North West has played an integral part in encouraging earth burns in Cumbria, with enthusiasts such as Brian Crawley ensuring that this knowledge is passed on to a new generation.

At the Weald and Downland Museum in Sussex, Alan Waters had been conducting burns since 1978, drawing on the knowledge of Mr and Mrs Langridge. Waters worked at the museum for twenty years, and was able to continue doing burns there even after he retired. Latterly he has continued the tradition at 'Charfest', a summer event held annually on the West Dean estate, timed to celebrate St Alexander's Day, the patron saint of charcoal burners.

In the Wyre forest Mr Nevey conducted a demonstration burn in 1973, and was able to pass on his knowledge when Lyn Armstrong was researching her book *Woodcolliers and Charcoal Burning*, in which she explores in depth the variations and detail of earth-burn technology.

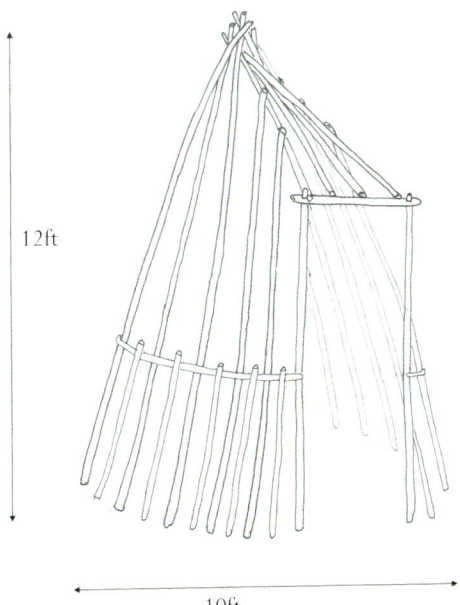

A new generation of charcoal burners: David Noblett and Rachael Jepsom.

Charcoal Burners' Huts

The huts that charcoal burners lived in are very particular, and are fairly similar throughout Britain. They consist of a temporary 'tent' or wigwam of long poles. The poles were often placed directly on the ground in a 10 or 12ft (3 or 3.7m) diameter circle, leaning into each other, some forked for the purpose at a height of 12 to 14ft (3.7 to 4.3m). The structure was covered in sacking and made watertight with turf or thatch. It may have had a stone fireplace, or just a brazier placed in the doorway to provide some heat. There is widespread archaeological evidence in Cumbria of woodland huts – dubbed 'bark peelers' huts', with a low circular or rounded oblong wall, and a flat stone hearth and chimney – which are also thought to have been used by charcoal burners.

12ft

10ft

Diagram of the pole structure of a charcoal burners' hut.

A large earth burn at the European Charcoal Burners Convention in Lembach, France.

The International Picture

Although the skills and knowledge required for charcoal burning in an earth clamp very nearly disappeared in Britain, there is a move to revive the skills and maintain this millennia-old tradition. This can also be seen in other European countries, with the work of the Europäischer Kolerverein (EKV) (European Charcoal Burners), a group first formed in 1997 in Germany, but which soon acquired members from Belgium, Switzerland, Italy and the Alsace in France. The EKV specializes in earth-burn technology, sharing knowledge and traditions through its website and biannual conventions.

Elsewhere in the world, particularly in Africa, earth burns are still a frequently preferred method, partly because of the minimal outlay required to set them up.

MAKING CHARCOAL IN AN EARTH KILN OR CLAMP

The Hearth, or 'Pitstead'

The hearth, also known as a pitstead, is the starting point for the building of the earth burn. The hearth should be level to allow for even stacking of the timber and to reduce the chance of the stack collapsing unevenly as the heap burns. When burning on a new site, or one which has not been used for many years, any vegetation and topsoil must be scraped away and retained in a large circle around the hearth for later use. It is important to ensure that there are no animal burrows or underground channels beneath the site, as these could allow oxygen in from underneath. If the site is steep, as many of the woods in Cumbria are, the hearth is dug into the banking and a retaining wall built up on the lower side to make sure the hearth is level.

Water Divining or Dowsing

Following a long career with the Shirley Aldred Company, designing cutting-edge retorts for developing countries, Don Kelley's technical knowledge of charcoal burning is unsurpassed. He stresses the importance of water divining to the charcoal burner. If you are unable to locate underground streams, you will not be able to pick a good place to site your charcoal kiln. Underground streams may make it impossible to get a good seal on a kiln or burn, in much the same way that rabbit holes allow the air in underneath. Also wet ground reduces the efficiency of the burn, having a cooling effect.

Don came and taught a master class in charcoal burning to the Coppice Association North West in the mid-nineties, and one of the skills he wanted us to learn was dowsing. He directed the aspiring dowser to take two metal rods (these can be made from metal coat-hangers) and to bend each into a right-angle to form a short handle and a long length. The rods are then held by the short handle, one in either hand, with both hands extended in front of the dowser so the long lengths of rod are about 9in (20cm) apart at right angles to the ground. Grasping the rods firmly, the dowser must concentrate on the ground in front, and walk slowly forwards across the site. If the sticks move of their own volition, twitching or even crossing over one another, this may indicate the presence of water.

Preparing the pitstead. COURTESY OF JOHN ALLONBY (JACK ALLONBY'S SON)

The final stack needs to be a true circle to maximize stability as it burns and shrinks. At the start of the build a peg is driven into the centre of the pitstead clearing, and an improvised compass of string is used to scribe a circle up to 30ft (10m) in diameter. The standard size (if there was one) was about 22ft (7m) across; the demonstration burn photographed was more like 10ft (3m).

Preparing the Timber

Ideally the wood is cut in winter when the sap is down and the leaves off, and left to dry in stacks in the wood for at least six months. This explains why burning was traditionally often done through summer and early autumn, with the added bonus of maximum daylight. However, when the demand for charcoal was high, burning was probably extended throughout the year.

To prepare the stack, wood is cut to even lengths of 2ft, 3ft or 4ft (60cm, 90cm or 120cm), with smaller-diameter wood ideally cut with a deft blow of the axe to create a sloping end that the log, or billet, will stand on, reducing contact with the ground and allowing

a little more air flow across the bottom of the kiln. Smaller lengths of wood (3ft/90cm) were known in the North West as 'coal wood', and the long lengths (4ft/120cm) as 'shanklins'. Diameters would range from 2–6in (6–15cm). These were the predominant sizes in coppices managed on a fifteen- to twenty-five-year cycle.

The lengths are taken by wheelbarrow from wood stacked in 'cords' at the side of the site ready for the burn. A cord is a system for measuring timber by stacking it in 4 × 4 × 8ft wood piles (1.2 × 1.2 × 2.4m). Alternatively it may be arranged in a circle just outside the circumference of the intended clamp so that the timber is easily to hand.

Tools Needed for the Earth Burn

The tools required for a burn are fairly minimal, the axe perhaps nowadays being replaced with a chainsaw. A long-handled shovel is very useful, as is an iron bar for making holes in the burn to allow in a controlled air flow and to 'feel' the progress of the burn. Traditionally one would add a 'sod drag' and a 'cooler', tools rarely seen these days. The sod drag is like a right-angled

A metal bucket, a charcoal riddling fork, a long-handled rake and various long-handled shovels, plus the screening behind.

A 'sod drag'. SIMON HANSON

Tools Required For Charcoal Production

The primary tools required for charcoal production in the study area (Mozambique) include: axe, hoe, rake and shovel. Ninety two percent of the male producers owned one or more tools compared with seventy one percent for female producers. The axe was the most common tool owned by all the producers owning one tool or more. The distribution of tools reveals a mean of two for both the number of tools owned and rented. Each tool was rented at a price of 50 Mts (US$ 2), but in some cases they were borrowed from family members for free. All of the producers with four tools rented out their tools for extra income.

From: Herd, A. R. C. 'Exploring the socio-economic role of Charcoal and the potential for sustainable Production in the Chicale Regulado, Mozambique', University of Edinburgh 2007, page 48.

fork or pitchfork for spearing the turves to pull them off the stack to allow water to be introduced at the end of the burn. A cooler – known by many names, such as rib shovel, rabil, rebel, rubler, rauble or wiper – is also a long-handled tool with a board set at a right-angle, used to draw the earth down and off the completed burn without disturbing the charcoal.

Before the Earth Burn Starts

The major consideration before the burn begins is, 'Is there enough water to hand?' 50gal (250ltr) of water may be needed for a large clamp. If there is not a water supply on site, water must be brought in with a bowser or similar container that can hold a minimum of 50gal. The old burners would carry the water in a barrel supported by a wooden frame called 'stangs', which made it possible for two burners to lift an 18gal (82ltr) barrel. They would need three of these to fetch water from the nearest water source.

The Chimney or Flue

There are several ways to start building an earth burn, but all involve creating a central cavity for starting the fire. This can be at ground level, but if the stack is double height it may be that the central section of the first tier of timber is solid so that the fire has to work its way down to the ground during the burn. For a more modest burn a triangular flue can be built to a height of 4ft (1.2m) using even-length logs about 2ft (60cm) long, placed one on top of another to create a sturdy structure, which the logs can be leant up against. This system is the one most commonly used in traditional earth burns in the South East of England.

Elsewhere, and particularly in Cumbria, a stout post known as a 'motty peg' was used in the centre of the earth burn; the logs are leaned against this post, which can be withdrawn once the stack is built in order to make enough space for lighting the stack. For a bigger burn the first tier is constructed without a flue, and the motty peg is inserted into the centre before the second tier is constructed around it. The motty

The flue of an earth burn.

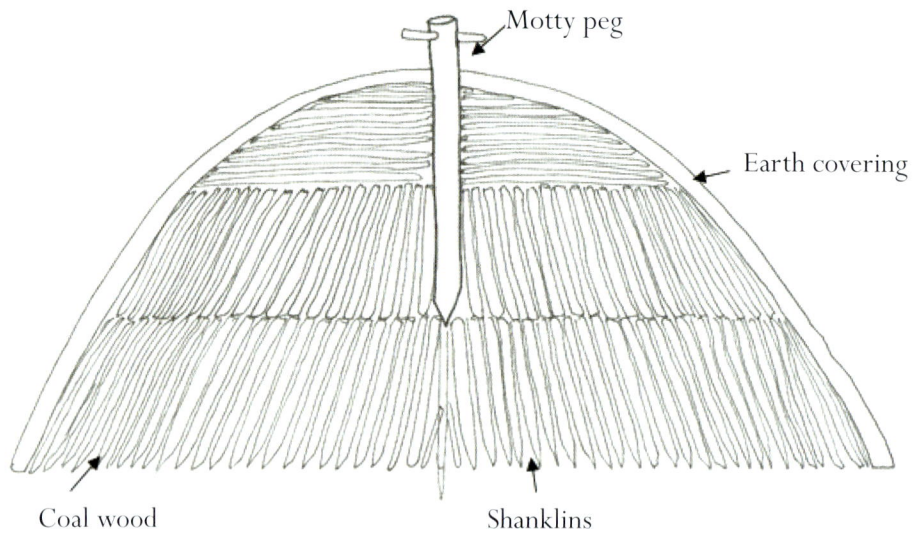

Motty peg

Earth covering

Coal wood

Shanklins

Diagram of the motty peg.

peg may have 'lugs' on it to help in removing it when the top tier is complete.

Alternative flue options include a tripod of logs lashed together to make a 'core', against which the timber is then placed; this would prevent the flame used to light the kiln going straight to the bottom, though it still needs a cavity above to accommodate the kindling. In Alsace, the method is to construct the burn around a metal pipe, which is withdrawn once the burn is lit. Other continental variations involve metal rings that hold the first three poles apart, allowing a channel to remain as the stack is built around it.

There are reports that a burn was sometimes lit from underneath. In this case a channel would be needed so as to be able to push the kindling and lighting coals through to the middle. Building the stack on top of a horizontal pole would be the trick here, but it would not be an easy job to remove this horizontal log prior to lighting the stack through the cavity created.

Building the Stack

4ft (1.2m) logs known as 'shanklins' are placed all round the flue in ever-widening circles, the

Building up the stack with 'shanklins' – 4ft (1.2m) logs of larger diameter.

Smaller-diameter wood – 'coalwood' – is added to the outside and the top.

larger logs towards the centre, and keeping the air gaps between the wood to a minimum. The logs are allowed to flare out slightly at the bottom to make a steeply sided but slightly conical stack, up to a diameter of 8ft (2.4m) across, or more for a bigger burn. A second or even third tier can be built at this point, before a cap of the thinner 'coalwood' is placed on top, keeping the centre hole clear.

Coalwood would also make up the final layer all around the stack, because heat will be lost towards the outside of the burn, and the thinner wood stands more chance of converting to charcoal, as the burn progresses to the final stages. Small pieces of wood are used to fill in gaps and to create a smooth, even surface to the stack.

Shingling the Stack

The stack is now covered with a layer of loose, dampened hay, straw, bracken or any weeds removed from the hearth. This provides a 'key' for the earth layer and stops the earth from dropping into the stack and contaminating the charcoal.

The Earth Covering

There are important variations of technique at this stage of the earth-burn build. When suf-

A loose covering of damp hay, grass, weeds or bracken.

ficient turf was available, the stack would be 'shingled' with slabs of turf arranged around from the bottom up and tightly packed, the grass side inwards. In other traditions, and perhaps when turf was scarce, the clamp was covered with a layer of loose, sifted soil or 'sammel'. Sammel may be a Cumbrian term and seems to refer specifically to a kind of sandy subsoil dug especially for the purpose, which is unlikely to be found in any quantity in the stony

The sifted soil or 'sammel'.

terrain of the Lake District woodlands. Another term used was 'mould', which Sam Hannah in his film reports 'is the work of women and children to dig'. The mould is then brought in swill baskets to the burn site.

Mr and Mrs Langridge, as reported in Lyn Armstrong's book, used the sifted debris from a previous burn, which would have consisted of the sandy mineral content of the soil when the humus had perhaps burnt away, and charcoal dust. They referred to it as 'gold dust', and it was bagged and saved for the next burn, even if they were moving sites.

An earth burn may have just a single cap of turves on the top where the heat is most intense. In Sam Hanna's 1973 film, Jack and Bill Allonby are filmed in Cumbria building a large three-tier charcoal clamp, the flat top of which they cover over with sods cut from the coppice nearby, carefully trodden down to create a tight seal on the vulnerable top surface of the stack. The 'mould' is then brought in swill baskets and

dumped around the top of the sloping sides and scattered down the sides of the stack until all the hay is covered.

If the mould was newly dug, then a deep subsoil was sought which had a high mineral content; clay was considered unsuitable, but was used if it was all that was available. At the burn shown on page 41, the over-enthusiastic application of soil was gently discouraged, as this would have a distinctly detrimental effect on charcoal quality. Instead, the burners were instructed by the charcoal master Alan Waters to 'let it breathe'.

Before Lighting the Kiln

The stack is now ready to light, but before a match is set to it a windbreak must be created around the kiln. This was traditionally 6ft (2m) high hazel hurdles, but could equally have been frames covered with sacking or brushwood such as birch-tops or besom. The wind can get up quite quickly, so particularly if the burn is likely to take a few days to complete, a wind-

RIGHT: *The clamp covered with soil and ready to light.*

BELOW: *A large burn in France with a thick cover of turf and soil.*

How to Make a Hurdle Screen (courtesy of Jack and Bill Allonby)

Two long poles are halved or flattened on one side. Five holes are drilled with an auger, evenly spaced, to take the hazel uprights and the frame assembled. Hazel rods are woven in and out to create an open hurdle. Bundles of bracken are then scattered generously over the screen, and further hazel rods woven in to create a bracken 'sandwich'. These panels were then propped up around the burn, and moved when the wind changed direction.

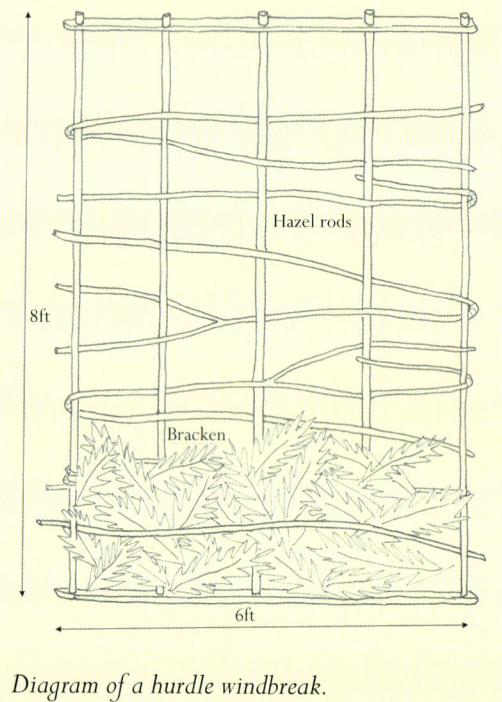

8ft

Hazel rods

Bracken

6ft

Diagram of a hurdle windbreak.

break prevents the burn being fanned from one side, when it can become either uncontrollable or unevenly burnt.

Lighting the Kiln

Either a channel has been maintained which leads down to the triangulated chimney, or the motty peg is removed from the centre of the stack to reveal a cavity 3ft (1m) deep and 5in (13cm) wide. Use a long shovel or a metal bucket to pour the burning coals down the flue, adding a little more charcoal or dry birch kindling, and giving it a tap down with a pole. In moments the stack will be alight and the flames will get going.

The Charfest burn in the photographs opposite and overleaf was on one of the warmest days of the year and quite windy. Within minutes the stack was alight and the central hole covered with green turf. The fire became very lively, and it took quite a few volunteers about twenty minutes of applying more wet hay and soil to the outside to bring the burn under control.

Watching the Burn

There is something strangely humbling about attending a burn. The earth 'clamp' is quite definitely 'alive', and the power and heat of the fire within is always attempting to break out. When flames break out, it is like a little window into an inferno.

As the fire is unpredictable, it is obvious why burners never worked alone, but in twos or threes or family groups. Someone must be vigilant at all times, ready with a shovel to shift

Alexander, Patron Saint of Charcoal Burners

Alexander, a Greek from Comana (in what is now Turkey) lived during the third century AD. After he converted to Christianity, he was said to be so pious – and good-looking – that he chose charcoal burning as an occupation, because he thought being covered in charcoal dust would disguise his features and allow him to live a humble life without drawing attention to himself. He was nevertheless discovered by St Gregory Thaumaturgus and made a bishop against all odds. He was well loved by his people, but ended his life as a martyr for his faith when Christians were persecuted under Emperor Decius, who decreed he should be burned alive.

Ger van Marion (GP) ceremonially lights the burn at Charfest.

Mark Cox and Peter Jameson cover the central hole with a green turf.

The burn can seem 'alive'.

The burn is getting rather too hot and needs further covering. AMANDA BINGLEY

The burn at night.

The burn nearly complete.

some of the earth 'shell' back up, when gravity has allowed the soil to trickle down the steep-sided stack. With more serious holes created by a sudden explosion of wood gas or where an unstable part of the stack has caved in, a bung of turf or occasionally tin, or wet hay and earth may be needed – but it is mainly a case of redistributing the soil back up the clamp. Night-time is perhaps the most special time, with the darkness amplifying the light from the glowing, smoking kiln.

At first light an early morning inspection reveals a shrunken but still animated heap, with a shell-like crust and little smoke.

Continental-Style Earth Burns

By contrast to the Charfest earth burn described above, the burn observed in Alsace, France, was considerably bigger. The larger burn illustrated opposite (bottom image) took 689sq ft (64sq m) of timber and measured 32ft (10m) across. The slightly smaller version shown below was about 18ft (6m) in diameter and would have taken 215sq ft (20sq m) of timber. A much thicker layer of turf covered the sides, with leaves and soil on top, which was trampled down into a solid covering. The fire, once started, was steady at first and the central cavity was soon covered with a block of wood and more turf. The smoke was only allowed to exit via four small holes at the top and four air holes at ground level. The burn seemed to be almost extinguished, but it is the norm to leave it to a slow burn, and by the next day there was sufficient smoke issuing to reassure us that the burn really was alight.

The Alsace burn was scheduled to last for eight to ten days, and the process was calmly tended by a rota of people working alone. The main task is to prod the clamp with the iron bar and listen to the sound it makes as it enters the stack. A thud of wood and the burner know it needs longer; a 'scrunch' of charcoal and the burn can be guided down the clamp by introducing new holes lower down the sides and shutting off the higher ones on the stack. In fact the earth covering, which hardly seems to need augmenting, is deliberately trampled down to exclude the air and encourage the burn to move down the stack.

The charcoal maiden lights the burn in Alsace traditional dress. LYNNE ETHERIDGE

The Alsace burn twelve hours from lighting.

The Alsace burn after twenty-four hours.

The Alsace burn, which was lit six days in advance of the event and was on day nine.

Introducing water at the top of the burn to create steam within, which will exclude any air.

Putting out the Fire

Quenching the fire is perhaps the greatest skill of an earth burn, as too much water will make the charcoal wet. The aim is to introduce just the right amount of water to create steam, which will exclude oxygen from the clamp, and in combination with a careful covering of fresh damp soil is sufficient to extinguish the burn. Then all that remains is to allow the heap to cool sufficiently so there is no re-ignition on exposure to the air. The turves on the top of the clamp can be raked off with a 'sod rake', and buckets of water poured over the hot coals. The turves are then replaced, thereby trapping the steam inside the clamp and, by extension, excluding all oxygen to starve the fire. If this process is done correctly the clamp can be left to cool over a few days.

Seemingly, the 'continental' burn observed in Alsace did not require heavy quenching, but was burnt to a completion and smothered with the soil to extinguish the fire. Just a little water was needed to sprinkle on any 'hot spots' that occurred.

Bagging up the Charcoal

When the charcoal is cooled and ready to bag, the layers of charcoal are exposed by raking off the soil covering, starting from the base of the clamp. The art is to be vigilant for any tell-tale tinkling sounds or signs of white ash forming, which indicate that the charcoal is re-igniting where 'hot spots' are exposed to oxygen. At this stage spontaneous combustion is not unusual. Cold, damp soil kept aside for the purpose is the best way to cover up lively areas (plus a sprinkle of water) – and more patience. The charcoal should retain the structure of the wood, but as you rake it out, it will break into pieces.

Charcoal Yields

Yields vary depending on the heat of the burn and on how green the wood was initially. Kiln burns (as opposed to retort burns) rely on burning some of the wood to create the heat which drives off the volatiles within the wood: this is known as sacrificial wood. More sacrificial wood is required to raise the temperature of wet or green wood. Getting the kiln efficient-

Raking out the charcoal.

Moisture Content and How It Affects Charcoal Yield

Theoretical reduction in charcoal yield from a kiln as a function of moisture content of wood charge.

From: *Charcoal Production — A Handbook* (Natural Resources Institute)

Per cent moisture	Per cent reduction in yield
56.1	17.4
46.8	11.9
44.8	11
39	8.7
28.1	5.3
16.7	2.7
9.1	1.4
4.8	0.7

The Size and Efficiency of the Kilns

In an interesting study of charcoal burning in Mozambique, Alastair R. C. Herd, an MSc student at the University of Edinburgh, examined the size and efficiency of the kilns. He was keen to gauge the potential for developing sustainable charcoal production for the increasing demand from a growing urban population, coupled with a problem of deforestation and degradation of habitats. The kilns he described were mainly earth burns on level ground, built in an oblong shape with the timber running across the width of the kiln.

Herd calculated that in order to keep up a sustainable supply of timber to provide for the current demand for charcoal, an area of 6,388 acres (2,585ha) of woodland would need to be managed on a coppice-with-standards system, cut on a thirty-year rotation.

The 'brown smoke' stage is altogether more painful for the eyes and chest.

ly up to a temperature of between 400–500°C and keeping this steady is a key to good yields.

The other factor is shutting the kiln down at the point just before the carbon is lost to combustion, or conversely avoiding shutting it down too soon, and ending up with masses of unconverted wood or part-converted wood, known as 'brown ends', 'parchar', or brands. A ratio of 1:6 is frequently cited as a ratio of wood weight to that of finished charcoal from a kiln. However, experience suggests that this is much more likely to be 1:8 or even less for an earth clamp. This is especially so when run as an experimental exercise, rather than a skill used frequently enough to become perfect – at least this is true for Britain. The table on page 47 shows the moisture content and how it affects charcoal yield.

A Word on Smoke Signals

A charcoal burner gets quite familiar with wood smoke, and earth burners more than most. Being in and around the clamp, ingesting a lungful of smoke is usual, especially in the early stages. Mercifully the high proportion of water vapour

of the early 'white smoke' stage does little harm – but soon it turns to a much more unpleasant 'brown smoke'.

Altogether more painful for eyes and chest, brown smoke contains the volatile oils, or naphtha, which give a characteristic bitter taste in the throat. Eventually the smoke becomes thin and wispy, the sign the burner is looking for, as it is an indicator of when the charcoal is ready and the kiln should be shut down. When the smoke turns blue, the time to close the kiln is very near, as it is a sign of the carbon burning. If left to burn, the result will eventually be less charcoal and more ash.

Conclusion

With the recent resurgence of interest in earth-burn kilns, several people in Britain are now proficient at conducting these burns. As with all aspects of charcoal making, there is much scope for disagreement about the 'best' way. One limiting factor is the sheer quantity of timber required for a full-size burn. Hence, many of the earth burns undertaken these days are small

The finished charcoal still steaming.

'demonstration' burns, which perhaps do not reflect the true potential of this fascinating skill. The European Charcoal Burners (EKV) undertake burns on a very large scale, and their members have a great deal of experience in large, slow burns. They have created a festival of charcoal burning to celebrate the skill and knowledge of their forebears. It is vital that these skills are preserved, especially at a time when more efficient and less polluting methods such as retorts are becoming more mainstream.

MAKING CHARCOAL IN PIT KILNS

Pit kilns are used very widely around the world, especially in West Africa. This is a specific method, as described by the Food and Agriculture Organisation of the United Nations (FAO), reporting on charcoal production in Guyana. There are many variations in the size of kilns created according to this specification – it is possible to make charcoal in a pit as small as a metre square.

How to Create a Pit Kiln

The large pit method requires a site with deep soil (ideally a sandy loam), which allows excavation of a pit 20ft (6m) long by 9ft (2.7m) wide. The depth is 4ft (1.2m) at the shallow end, extending to a depth of 8ft (2.4m) at the far end.

Loading a Pit Kiln

First some logs are cut, five to fit across the pit and four to run the length of the kiln. This creates channels for the gases to pass under the mass of timber and burn from the bottom up, within the pit. The pit is then filled with timber cut to the width of the pit – in this case 9ft (2.7m) – and stacked fairly tightly, up to above ground level, to maximize production. A channel is left at each end to act as an inlet and outlet port.

Creating a seal: The timber stack is first covered with some vegetation 'brush' to create a 'cap', which allows the earth to be placed on top without falling through into the wood stack.

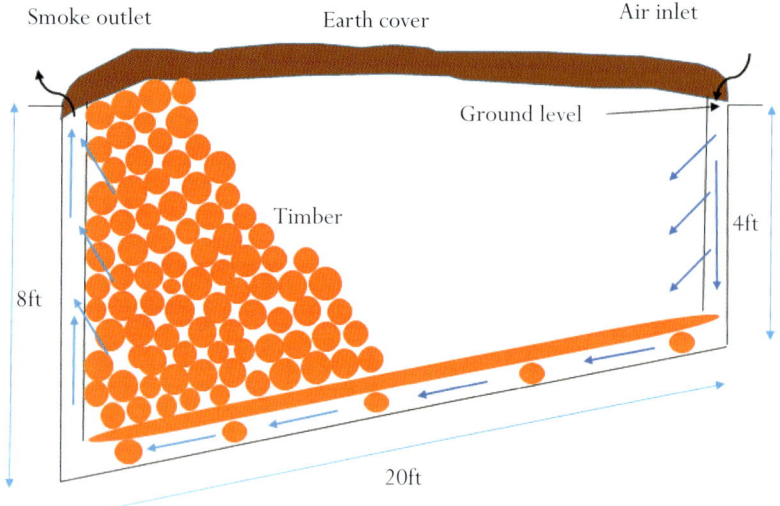

Diagram of a pit kiln.

Smoke outlet   Earth cover   Air inlet

Ground level

Timber

8ft   4ft

20ft

A layer of sand or sieved soil is then added to a depth of 12in (30cm).

Lighting the pit kiln: Using some kindling, the kiln is lit from the shallow end, adding part-burnt charcoal to get the fire going. The burn progresses from the shallow end towards the deep end. Amazingly, it can take twenty to thirty days for all the wood in the pit to carbonize. During that time the earth 'crust' on top must be kept intact, even as the wood sinks down in the pit. Constant vigilance is needed to patch up any cracks. Even so, the wood at the shallow end will inevitably be more liable to burn to ash, resulting in losses in production.

The effect of rain: One of the drawbacks of such a long burn is that rain can penetrate the clamp, washing the soluble volatiles back into the charcoal. The composition of these soluble volatiles is largely acetic acid, which if it contaminates the charcoal can destroy the bags that are used for storage.

Shutting down the pit kiln: When the burn is complete – mainly gauged by the loss in volume – the pit kiln inlet and outlet must be sealed, together with any further gaps or cracks. The fire is thus essentially smothered through the complete exclusion of air. Some pit kilns include the modification of sheets of tin to help the sealing process.

Cooling the charcoal: The pit kiln can take as long as forty days to be cold enough to excavate the charcoal. As with any burn, the danger of re-ignition is always present, and a supply of water for quenching is vital.

The comparative merits of pit kilns: It is possible to make a great deal of charcoal in a pit kiln with little or no outlay, beyond the labour involved in digging the pit and looking after the burn. The quality of the burn and the efficiency of conversion is very variable, and the skill of the burners is vital to ensure a decent return for effort. It is really only viable where ground conditions are very favourable, wood supply is abundant and low cost, and where skilled labour is cheap and readily available.

Issues with Earth Burns and Pit Kilns

The efficiency of earth-burn or pit-kiln technology is very variable, and several factors may impact on it: the skill of the workforce, the quality of the material to be converted to charcoal, and the moisture content of the feedstock.

Skill of the Workforce

Where there is continuity of charcoal burners in families and through apprenticeships, skills are passed on and essential knowledge is re-

Traditional burn very nearly complete in France.

The inside of an earth burn.

tained over the generations. In these instances, the highest level of efficiency is expected. However, this can be counter-effective where adherence to tradition stifles innovation and potential improvements.

Quality of the Material to be Converted to Charcoal

Uniformity of the feedstock makes for more efficient stacking of timber for most methods of charcoal burning, but it is particularly important with earth burns that logs of uniform shape and thickness are used to make an even, stable burn.

Moisture Content of the Feedstock

The main reason for low yields in earth burns or pit kilns is the timber's water content. Where demand is high for charcoal, and there are accompanying economic pressures, there is a tendency to burn timber that is cut 'green'. This results in a much higher loss of 'sacrificial wood' to drive out the excess moisture.

It is likely, however, that these types of kiln will continue to be used wherever the capital costs of more modern equipment are prohibitive. The most important issue facing the charcoal industry worldwide is managing a truly sustainable forest resource, whilst maintaining protection for fragile ecosystems and species diversity, despite exploiting what is a renewable resource to meet essential fuel needs.

A comparative table for different ways to make charcoal is found in the Appendix.

Chapter 4
Metal Kilns and Brick Kilns

PORTABLE METAL KILNS

The metal ring kiln has been the main type of kiln that small-scale British producers have used to make charcoal since the mid-twentieth century. Metal ring kilns have obvious advantages over the earth kiln. Firstly, it is possible to get a really good seal on the kiln, and thus to regulate the flow of air in and gases out through a system of chimneys and vents. Secondly, once the kiln has reached the correct working temperature and the lid is down and sealed it can be left to burn, leaving the charcoal burner free to do other things – including getting a good night's sleep!

The disadvantage of the metal kiln is that, unlike an earth burn, the lack of insulation around the kiln surface allows heat to radiate out into the atmosphere, thereby requiring more sacrificial burning of the timber feedstock to keep the kiln at a working temperature. The disadvantage of metal ring kilns over retort kilns (*see* Chapter 5) is the large amount of smoke containing gases and volatiles that discharge into the atmosphere.

Two people can relatively easily operate two 7–8ft (210–240cm) metal ring kilns. Filling, burning, emptying, grading and bagging over a twenty-four hour period will produce, from a kiln this size, approximately 1,100lb (500kg) of charcoal. The kiln needs to cool for twenty-four hours before starting the cycle again, and in hot weather it can be even longer. Realistically, a maximum of two burns over a six-day working week can be achieved. To increase the capacity of the kiln, a second tier can be added, thereby obtaining more charcoal from a double-decker kiln.

A recent development has been open kilns for biochar production. The Kon-tiki kiln

A double-decker kiln in action.

developed by the Ithaka institute in Switzerland is an example of this technology. These kilns work using what is known as 'flame curtain pyrolysis', where the burning gases shield the spent feedstock from oxygen, preventing the carbon from turning to ash.

On a smaller scale, a 45gal (205ltr) metal drum or barrel can be used to make charcoal very cheaply and easily. This method has the advantage that old metal drums can be obtained practically free. It is possible for one person to run eight or ten barrels and produce up to 220lb (100kg) of charcoal in a working day, as described below.

Barrel Burns

Obtaining a Barrel

Metal barrels or 45gal (205ltr) drums can be obtained from agricultural suppliers, motor factors or scrap-metal merchants, or on line if you don't mind paying for them. Ideally the metal barrel will have contained a non-flammable and non-toxic substance, such as molasses. It is crucial that the barrel has not contained any volatile substance such as petrol or paraffin, as health and safety could be seriously endangered when sparks fly while cutting the hole in the side. The barrel will probably be painted, and as the paint is burnt off during the first burn it will give off unpleasant smoke. It is also vital to check whether the bung is made of plastic or metal: a plastic bung will burn away, and the bung hole can be sealed with a clod of earth.

Cutting the Slot in the Barrel

A barrel for charcoal burning can be placed either upright, or lain down on its side. The horizontal position has the advantage that it is easy to continuously feed the barrel until it is full of charcoal, and to seal it when the burn is complete. Then all the burner needs to do is roll the barrel over so that the slot is underneath, and seal around with soil (*see* page 56). The slot along the side is large enough to allow sufficient air in to get the burn up to temperature, and can be tilted into the wind, as required. In 1993, the author was shown this type of barrel charcoal burn by Richard Edwards, who set up the original Coppice Association.

The upright version is a little trickier as it requires sufficient air-holes to get the fire going in the bottom of the barrel, which are harder to seal, and the lid may present a challenge when the time comes to shut it down.

Cutting a slot in the side of the barrel.

To make a 'mini' barrel kiln, lie the barrel on its side and use an angle grinder to cut two slots 7in (18cm) apart, then cut along the short ends to create a 'letter-box' the length of the barrel. Always wear gloves and safety glasses for this: the sparks fly alarmingly, but are harmless as long as the cutting is done sensibly, and the cutter, and anyone close by watching, wears eye and skin protection. The cut edges are then folded back so that no sharp metal edges are exposed, which could cause injury when the operator reaches in to load or unload the barrel. The finished slot should be 9in (23cm) across. It is possible to do this operation with a coal chisel and lump hammer, but it is hard work and noisy, and much more difficult to get a finish that will not be uneven and sharp, and risk cutting the burner's hands.

Having made one lovely barrel kiln it may be tempting to make another, as it is just as easy to run two at once (realistically, one person can run up to eight barrels). A great advantage of the barrel burn is that the barrel kiln can easily be taken to the wood stacks. A barrel can be carried a fair distance on one shoulder or with a friend to carry an end, or transported in a wheelbarrow.

Importantly, choose level ground for the burn, with a decent layer of topsoil (not peat, which may ignite). Dig out a shallow trench just longer than the barrel and about two spades wide, and keep the soil handy for later. The barrel kiln will sit comfortably in this trough or just to one side, using turf or a chunky log to wedge it upright. If grass turves have been removed, keep them intact because they can either be replaced on the site later, leaving minimal disturbance, or can be used as stoppers to block any air holes.

Lighting the Barrel Kiln

It is really useful to have a bit of charcoal to get a burn started. This is not absolutely necessary – and probably not worth going to a garage to buy imported charcoal for this purpose – just remember when emptying the barrel to put some aside for the next burn. If no charcoal is to hand, have plenty of dry kindling and dead wood

Start two fires at the bottom of the barrel.

Fill the barrel with logs cut to half-barrel length.

ready. Eco firelighters are brilliant for lighting, as ideally the two fires are set equidistant from each other and from each end of the barrel, to ensure a good spread of fire at the bottom of the barrel. Newspaper is also fine, though it may be difficult to keep paper dry in the woods.

Feed the fires first with small sticks, gradually using bigger sticks as it gets going. It is important not to rush this stage, as the aim is to get a steady build-up of heat, so it is crucial to keep adding dry kindling regularly to get the kiln up to temperature. If the fire is too slow and short of oxygen, tip the barrel mouth a little to pick up any breeze, and if there is no wind, use a handy 'wafter'. This first stage is the equivalent of the 'free burn', and ideally it should take about an hour to achieve a glowing bed of coals across the bottom of the barrel, about 4in (10cm) deep.

Loading the Barrel Kiln

The barrel is now ready to fill with logs, which should be pre-prepared by cutting to approximately 16in (40cm) long, so that two logs laid lengthways take up the full length of the barrel. Any that are over 4in (10cm) in diameter should be split. Once the barrel is packed as full as possible with logs, for a while white smoke (steam) will issue from the barrel. Then as the wood heats and the volatiles start to be driven out, the smoke will become more acrid and brown.

The brown-smoke stage is the main disadvantage of barrel burns, as it is quite difficult to tend the barrels (especially if doing more than one) and avoid the smoke. Inhaling the smoke is inevitable, but should be kept to a minimum, as it is toxic and can cause headaches. Eventually, usually a further hour into the burn, flames will start to break out once the wood has shrunk and settled in the barrel. Now more logs can be added, piling them up in the mouth of the kiln to exclude the oxygen, to slow down the burn again. Use gauntlet gloves for this as the flames can lick up unexpectedly and singe skin or clothes. This process can be repeated until the barrel is so full of glowing coals (charcoal) that the flames are hard to suppress: it is now time to shut down the kiln.

Shutting Down the Barrel

Be prepared for this moment by wearing a thick pair of gauntlet gloves. Grasping the barrel firmly on each side, walk it off the immediate burn area until it is in a position judged to be the right distance from the trough dug earlier so that the slot (which is still at the top) can be positioned in the centre of the trough when it is underneath. Now, unless the barrel is flipped over really quickly, there is a risk that the charcoal will tumble out as the barrel is turned. To avoid this, place a few short logs crossways inside the barrel and caught under the rim on each side. This will help prevent the wood from falling out. If a little escapes, it is not a major problem as it will hopefully be contained in the trough, which can be filled in around the barrel to stop the smoke from escaping and hence close any holes where oxygen can get in.

It takes around five to ten minutes to go round the barrel firming down the soil, adding more where needed until it is certain that nothing can escape. If the soil used is damp there will be some steam wherever it comes into contact with the hot barrel, but this will soon dry off.

The barrel will cool surprisingly quickly, and can be touched once the fire is out and the heat starts to dissipate, which will happen as soon as the oxygen is excluded. It must be quite cool before risking rolling the barrel back to look at the charcoal spoils. Leave it for at least two hours; even then it is advisable to have water ready for sprinkling, if there is any sign of a danger of re-ignition. It is best to leave the kiln overnight: by the morning the barrel will be stone cold and ready to unload.

Unloading the Barrel Kiln

Roll back the barrel to expose the barrel slot, and pick up and bag any charcoal that has come out into the trench. Some of the wood may have only partly turned, or not turned to charcoal at all. This 'par char', or 'brown ends', should be removed and stacked to one side to be used for the next burn (once one burn is happily achieved, there is always an incentive to have another). The wood least likely to have converted to charcoal will be found at the ends of the barrel – which is why it is important to keep the fire burning evenly along the full length of the barrel.

The barrel is left to cool.

When all that remains is lovely charcoal, take a spade or shovel and scoop the charcoal out into bags. It will need 'grading' into different sizes, so don't worry if some of the charcoal is in quite fine particles or just ash. Commonly, in barrel burns the charcoal particle size is smaller than that produced in a larger kiln – and much smaller than charcoal from a retort. The charcoal is smaller because it is converted at a higher temperature and the structure of the wood is more likely to shatter. This is certainly the case with wetter wood, which is more inclined to shatter as the moisture is forced out. Thus, the greatest effect on particle size is the temperature and how fast the burn progresses. The process of grading the size and bagging the finished charcoal is described in detail in Chapter 6.

Metal Ring Kilns

Metal ring kilns are portable and quite easily set up and moved from site to site. Like the barrels, they can be taken to the timber resource to reduce the transportation of heavy wood. These kilns are also brilliantly easy to fill when used in conjunction with a firewood processing ma-

Roll back the barrel to see how much charcoal the burn has created.

Using a firewood processor to fill the kiln. MIKE CARSWELL

Biochar in a Barrel

An alternative version of the barrel burn method is to treat the barrel kiln more like the Kon-tiki kiln and use a technique known as 'flame curtain pyrolysis' (*see* the description of a Kon-tiki kiln later in this chapter). In this process, wood is added layer by layer and allowed to burn freely. The top surface of the burning wood in the barrel eventually oxidizes and shows a thin layer of grey ash, and at this point more wood is added. The rising currents from the hot gases inside the kiln draw air from up the outside of the kiln and a vortex is created, diverting the gases back down into the barrel, making the burn almost smoke free. A 'wafter' may be needed, to keep the oxygen levels up in the barrel and the gases ignited.

The way the wood burns in the open barrel is described as a 'flame curtain', shielding the embers underneath from oxygen and from burning away to ash. Instead the embers remain in a static state, their mass increasing as more wood is fed in. Eventually the barrel is so full and hot that it is impractical to continue and the kiln can be shut down in the conventional way. An alternative way to shut down is shown in the top photograph on the opposite page (*see* right), which has the advantage of preventing the biochar falling out as the barrel is turned over.

The burn in this photograph was only run for three hours. The feedstock was dry brash, and the yield 18lb (8kg) of biochar.

Burn the barrel hot and create fewer emissions.

A plastic lid can be used to waft and create a draught.

chine, which can cut the wood to length, split it as necessary and move it on a conveyor belt directly into the kiln. This process reduces the sheer effort and physicality of picking up the wood to load the kiln, and the fewer times the wood is handled, the more efficient and cost effective the system. However, it is not always convenient to take the wood processor to the woods, and there may be security issues. The kiln/processor combination often requires a yard where sufficient supplies of timber can be stacked, ideally with a dry space to store finished charcoal.

Obtaining a Metal Ring Kiln

Metal ring kilns can vary from 4ft (120cm) up to 8ft (240cm) in diameter. Larger dimensions make the lid too difficult to manoeuvre. A good-size kiln for one person to manage is 6–7ft (180–210cm) in diameter – but regardless of its size, emptying a kiln is always easier with two people. Metal ring kilns can be bought new; suppliers are listed at the end of the book.

Setting Up the Kiln

Choose a level area at least 12ft (360cm) wide for an 8ft (240cm) kiln, and 7ft (210cm) wide for a small 4ft (120cm) kiln.

Using turves to seal a barrel kiln is a useful method when your feedstock is twiggy.

The finished 18lb (8kg) bag of biochar.

Setting up the ports on a 4ft (1.2m) kiln. MARTIN HALE

Walter Lloyd and his Legacy

When Walter Lloyd came to the Lake District in 1986, he brought his skills of charcoal burning learnt during the war when he was evacuated to the Welsh borders and worked with charcoal burners on the banks of the River Severn. The Lake District National Park Authority were concerned that coppicing had ceased to be a viable business and that the woods were in danger of losing their diversity. Walter's son Bill Lloyd had set up 'The New Woodsmanship Trust' to attempt to reverse this decline; Walter suggested that charcoal making would be a good way to renovate derelict coppice woods, and was determined to prove that it would work.

His training had been with metal ring kilns, so he bought four 8ft (2.4m) kilns and went into serious production. He was famous for his perhaps rather optimistic predictions of how much money could be made from charcoal, and was responsible for inspiring many others to follow his lead. Running four kilns alone was an impressive feat, and there weren't many who had his energy or could match that output.

By the early nineties he had stepped back a little, and though he continued to burn charcoal, he had a mutually beneficial arrangement with local coppice merchant Bill Hogarth, who would bring the wood, cut to size, in his tractor and trailer and offload the logs straight into the kiln. He marketed the charcoal in polythene bags printed with the iconic brand 'Lakeland Charcoal'. These bags can still be seen in local garages with charcoal made by Walter's stepson Tom Barron. Walter continued to be an inspiration to the coppice and charcoal world in his role as President of the Coppice Association North West.

Walter Lloyd hard at work emptying one of his four metal ring kilns. DAVID JONES

Clear the site of vegetation, saving and pushing any top-soil towards the edges of the circle. Place the ports equidistant in a circle (there will be eight – or perhaps four for a smaller kiln.) The ports must be level so that the metal ring when placed on the ports makes equal contact with them and does not rock. Check that the ports protrude equally at least 4in (10 cm) into the kiln.

When a site has been used repeatedly for burning, it can become hollow from enthusiastic shovelling. This can be a problem if, during the time between burning and emptying the kiln, rain water collects and pools in the bottom of the kiln. This will result in wet charcoal, which will need drying, which is never easy. Either cover the kilns when cold with a tarpaulin, or lift the metal ring and level the site, creating drainage channels to divert rain away.

When the kiln is level and the ports all evenly spaced, the soil should be drawn back round the

The ports should all extend equally into the kiln by at least 4in (10cm).

The ring has been placed on to the ports.

The soil has been drawn up round the bottom of the ring.

Chopping the logs into halves.

ring to seal the gap, so the only air getting in is through the ports.

On some sites there is not enough soil available, and more will need to be brought in from elsewhere. Before starting to fill the kiln, check there are no rabbit holes or underground channels which could allow air in from underneath, as these are impossible to seal at the end of the burn.

Timber for Charcoal

Chapter 2 identified the various species of timber suitable for charcoal making. In this chapter the concern is how to prepare the timber for the kiln. As always, the principle is to reduce the handling of it – that is, the time and effort spent filling the kiln. Experience suggests that splitting the wood is not really necessary, however, when the timber is over 10in (25cm) in diameter, though some burners like to halve the log to ensure that all the wood carbonizes at the same speed.

An alternative to halving the logs is to cut them shorter, so there is less distance for the heated volatiles to travel to escape, and less likelihood of the wood shattering in the charcoal conversion process.

Cutting the Wood to Length

Timber can be cut to any convenient length for handling and fitting into the kiln. A smaller diameter kiln will need smaller pieces to fill the gaps and increase the volume of wood stacked into the kiln. A rule of thumb would be 4ft (120cm) long for 2–4in (3–12cm) diameter wood, 2ft (60cm) long for 4–6in (12–15cm) wood and down to 6in (15cm) long for 10in (25cm) and above timber.

Wood Machines

When considering investing in a wood processing or firewood machine there are many types to choose from. The wood machine is essentially a conveyor belt which loads the cut logs into the kiln. Choosing the machine with either a circular saw blade or a chainsaw blade may depend on whether there is a use for the sawdust: a chainsaw generates considerably more waste than a circular saw. Will it stand alone

with its own engine, or run off a tractor power take-off (PTO)? Experience suggests that a tractor-mounted processor ties up the tractor to one task, and is cumbersome to attach and detach. This may mean the machinery is kept mounted for the entire season, making the tractor unavailable for other jobs. Investing in a processor will depend on how much charcoal production is planned, or whether firewood deliveries may keep the machine busy, justifying the expense over the dark winter months.

The Initial Hearth or Platform

Whatever mechanical or manual method is employed to get the timber into the metal ring kiln, it is difficult to avoid jumping in and arranging the wood at the start, in order to create a platform on the base of the kiln.

An 8ft (2.4m) kiln needs sixteen fairly chunky logs to line each side of every port pointing into the centre of the kiln like the spokes of a wheel. The logs should be at least 4in (12cm) thick and 30in (80cm) long. Then use a range of different lengths of logs placed crosswise to form a channel that leads from the port to the middle, and fill in the gaps with more logs.

Making channels at the base of the kiln for air flow.

The platform of logs arranged on the bottom of the kiln.

Kindling and brown ends used to create a flammable core at the centre bottom of the kiln.

Kindling

It is important to use sufficient kindling to ease lighting of the kiln. Page 66 shows hazel offcuts from hurdle making being used, but any dead and dry brash will do. Use brown ends or par char rescued from a previous burn, wherever possible, which is perfect for starting a new burn. Two or three sacks of brown ends, placed as close to the centre as possible, should suffice.

Loading the Kiln

Fill the kiln until the wood stands up above the rim by at least 12in (30cm), to ensure that ample oxygen can be drawn in under the lid for the first 'free burn', as the burning wood brings the kiln up to temperature. If the lid comes down to rest too soon, the kiln never gets hot enough to produce the wood gas needed to fuel the burn. The picture above right shows the logs placed to hold the lid open on a burn where there was insufficient timber to stack up high.

Stacking the Kiln

There are many differing views on how to stack a kiln. Clearly, the more care that is taken packing the work, the more that can be squeezed in,

though be careful not to pack it too tightly as this can affect the overall burn quality. Some space is required to accommodate the limited oxygen needed for the wood to burn.

There is also the random large log approach: although the kiln in the picture (*see* opposite page, top) was filled by hand, it could easily have been loaded using a tractor front-end loader (*see* opposite page, bottom).

The benefits of loading with a tractor front-end loader are speed, and physically preserving one's back (the enthusiasm for bending and lifting logs up into a kiln can soon wane!).

Every burner will develop his or her own way of filling their kiln, finding an approach that increases yields/quality and profits.

Lighting the Kiln

Once the lid has been lifted into place, the kiln can be lit. The best method is to check that a straight hazel rod or similar can reach through the port with a clear passage to the centre of the kiln. Tie a rag around the end of the stick and douse it with a little paraffin – do not use petrol or more volatile substances.

The burning rag is then pushed through the port and slowly eased into the kiln to ignite the

The kiln filled with large logs to up above the rim.

Filling the kiln with a front-end loader.

ABOVE *A kiln stacked carefully (with more chopping).*

LEFT *Jamie and Sam lighting the rag soaked in paraffin.*

The 'free burn'.

kindling stack at the centre. Just occasionally it may need to be repeated if the kindling does not catch first time. Usually within a few moments some crackling is heard and the first puffs of smoke appear.

The Free Burn

The first hour or so is a 'free burn', where the kiln lid is kept open and the ports are all open wide. Masses of white smoke are driven off at this stage, which is mainly steam, as water is the first substance driven out as the wood begins to heat.

Soon, as the volatile oils are driven out, the smoke becomes acrid, catching the back of the throat at each breath. After an hour or so the lid is nearly down as the wood has shrunk and settled, though occasionally a stubborn lump of wood holds the lid up longer than might be preferred. The sign that the kiln is getting up to temperature is the flaring of the wood gas as it ignites, sometimes around the rim, sometimes

shooting out of one of the ports, which may catch your ankles.

In the photograph above a stick is used to keep the lid up a little longer as it does not seem to be quite hot enough. Overleaf, the chimneys are pre-warmed to help the smoke draw upwards when they are put in place.

Putting on the Chimneys

Every other port will become a chimney and will need the open end blocking with a brick and some soil. Alternate ports remain open to provide sufficient air to keep the burn alive.

On page 69 the picture above the chimneys have just been put in place – the further kiln had the chimneys pre-warmed. The kiln with the warmed chimneys is drawing better from the start, though the nearer kiln will soon catch up.

What Next? Put Your Feet Up!

This whole process for two kilns can be done with two people in four or five hours. By then it

LEFT *The first smoke is mainly steam.*

BELOW *Warming the chimneys just before putting them on the ports.*

The chimneys are now in place and sealed at the bottom, and the four air inlets are open.

is probably time to head home for a shower and a hot meal. One approach to charcoal burning is to extend that break into an evening watching TV and a solid eight hours sleep, by which time – fourteen hours later – the chimneys can be changed. However, other charcoal burners insist a more efficient burn occurs if chimneys are changed around regularly every four hours through the night. If you are hoping to get planning permission for a temporary dwelling to stay with your kilns overnight, far be from me to suggest that this is unnecessary. Certainly, moving the chimneys around allows a more even burn, with less ash where the wood near the inlet port has burnt away, and potentially a better yield. It

also means an eye can be kept on 'hot spots', and a port inlet can be semi blocked (particularly on the windward side) to keep the temperature down and create less heat stress on the kiln.

Shutting Down the Kiln

At some point on day two the kiln will need to be shut down – although this timescale will vary from kiln to kiln. Especially in a small kiln the burn length may only be some eight hours (in these circumstances postpone lighting the kiln until the early morning). The 8ft kilns pictured above – lit at 14:00 the previous day – were ready to shut down at 16.30 on day two.

The smoke is now wispy and blue, and the kiln is ready to shut down.

A sock 'rabbit' filled with sand, used to block off the air inlets and chimney ports.

The picture above shows blue wispy smoke with no energy in it. During this stage in the burn the chimneys were removed and replaced with old socks or 'rabbits' (*see* left), filled with sand that fit perfectly in the round ports, with brick and soil sealing the square holes. This is repeated on all the inlet ports, and fresh sand is added to the lid to suppress any leaking smoke, a sign that air is still getting into the kiln and re-igniting the burn.

Cooling the Kilns

Once the kiln is fully shut down, there is nothing to do but wait until it is completely cold. Semi-warm kilns can re-ignite, and the burn can be very hard to put out again once oxygen is introduced and it starts to crackle. Make sure there is a decent water supply available to quench the fire if this happens, as the entire burn can be lost.

Unloading

Emptying a kiln involves jumping in to shovel out the charcoal. A plastic shovel is much lighter than a metal one – and avoid trampling on the charcoal more than is absolutely necessary. Ideally shovel directly into a grading tool or 'riddle', for which process two people are an advantage. If working alone, it will be necessary to bag the charcoal into sacks and put it through the grader later, as it is physically exhausting to be jumping in and out of the kiln many, many times. Some burners avoid the grader process by using a charcoal fork that leaves the small particles behind, allowing the charcoal to be shovelled directly into bags. An efficient system avoids handling the charcoal too many times.

Ideally the charcoal is emptied directly through a grader and straight into the final bag in a seamless operation, keeping the bags clean and dry throughout the process. However, even the best laid plans can be ruined by a sudden rain shower. In these circumstances it may be better to put the graded charcoal into sacks and bag it later under cover. If the kiln has to be emptied on a wet day, consider placing a large tarpaulin over the kiln. (*See* Chapter 6 for a detailed description of bagging.)

The Viper Kiln

David Hutchinson of Yorkshire Charcoal developed a type of mobile kiln he dubbed the 'Viper', in a nod to the old custom, reported by Arthur Ransome, of charcoal burners keeping an adder in a box. He has refined the kiln over the years and it is still in production. It runs as a continuous process, which David describes as a 'cross-draft gasifier'. This is not dissimilar to a large-scale version of the barrel burn already described above, when the barrel is allowed to run hot, recycling the gases. A hot fire of charcoal from waste and brown ends is started in the chamber, then one tonne of timber is placed inside, and the process begins with a good deal of steam and white smoke. Once the wood begins to produce gas, it is circulated and burnt within the chamber.

The ingenious design allows the charcoal to be released out of the bottom of the kiln and into sealed drums for cooling and storing, making space for a new charge of timber to be added to the chamber. The kiln, which can be kept going indefinitely, is capable of producing five tonnes of charcoal a week at a peak conversion rate (with seasoned alder) of 4.3:1.

The Kon-Tiki Kiln

The Kon-tiki kiln was developed to convert brash and small-diameter wood to charcoal for incorporating in biochar. It was inspired by the cone-shaped kilns found in South America, which may have been the source of the charcoal element of the 'terra preta', or the fertile soils of the Amazon basin, which fed the early civilizations in that area. The kiln uses the technique known as 'flame curtain pyrolysis', described above.

The Kon-tiki starts with a fire at the bottom of the cone-shaped kiln, fed with dry kindling.

Once the fire is going well and is sufficiently hot, brash is added cut up into 2ft (60cm) lengths, layer by layer, and allowed to burn freely. The rising currents from the hot gases inside the kiln draw air up the outside of the kiln where the air is pre-warmed. A baffle around the edge encourages the air back around and

Robert Taylor demonstrates the Kon-tiki kiln.

The Kon-tiki Kiln in Britain

The Kon-tiki kiln technique also has a history in Britain. John Evelyn in his landmark book Sylva, first published in 1664, describes the process of making 'small coals' thus:

Made of spray or brushwood which is stripped off from the branches of copp'se wood, setting one of the bavins on fire two men stand ready to throw on bavin upon bavin. .. so they have burnt all that lies near the place to the number of five or six hundred, but ere they begin to set fire they fill great tubs or vessels with water and this they dash on with a great scoop , so soon as ever they have thrown on all their bavins, continually plying the great heaps of glowing coals which gives a sudden stop to the fury of the fire while with a great rake they lay and spread it abroad and ply their casting of water still on the coals which are now perpetually turn'd by two men with great shovels a third throwing on the water: This they continue till no more Fire appears, tho they cease not from being very hot: After this they shovel them up into great heaps and when they are thoroughly cold, put them up in sacks for London where they use them among divers artificers, both to kindle greater fires, and to temper and anneal their several works.

Ruth Thompson and Robert Taylor fit the baffle to the kiln.

Light a fire with kindling in the bottom of your kiln.

Developing a Horizontal Gas-Air Vortex

Once the kiln reaches its working temperature of 650°–700°C, hardly any smoke is visible. The combustion air rolls in over the metal edge of the outer wall and into the kiln. The burning gases must also escape upwards so, similar to a clockwork motion, a counter-rotating vortex is established in the centre of the kiln. Thanks to the establishment of this horizontal vortex, the air supply to the fire zone is stabilised. The wood gas, which is heavier than air, is kept in the vortex until it is completely burned. Thus, the second fundamental principle of Kon-tiki craft is the development of a horizontal gas-air vortex, which provides a stable, smokeless combustion regime.

From: 'Kon-Tiki flame curtain pyrolysis for the democratisation of biochar production', by Hans-Peter Schmidt and Paul Taylor.

A layer of white ash forms on the burning coals.

A new charge of feedstock gasifies immediately in the intense heat.

down into the kiln, creating a doughnut-shaped vortex, which in turn diverts the gases back down into the kiln, making the burn become almost smoke-free.

The top surface of the burning wood in the barrel eventually oxidizes, producing a thin layer of grey ash.

At this point more wood is added, and the vortex produced creates a 'flame curtain' which, similarly to the pit kiln or barrel burn, protects the embers underneath, so they do not burn away to ash but remain in a static state, adding to the mass as wood continues to be fed into the kiln.

It is possible to continue to add wood until it is about 4in (10cm) short of the rim of the kiln – any more prevents the flame curtain from functioning. A full kiln should yield 1.3cu yd (1cu m) of charcoal, with an approximate dry weight of 660lb (300kg). The burner can stop adding wood at any time, though it takes skill and nerve to leave it to the point where everything is converted to charcoal but has not started to convert to ash. A final layer of small chip or even sawdust can be useful, as it will carbonize quickly in the heat and reduce the part-burnt charcoal or brown ends in the final layer.

Dousing the Kon-tiki Kiln

The only way to put out the Kon-tiki kiln is to douse it with water. A considerable amount of water may be required to put it out completely, taking care as the heat in the kiln will initially produce a great deal of super-charged steam. This is similar to the process of charcoal activation, though the temperatures are not quite as high as they would be for maximum surface area production. It is advisable to wear gauntlets and a protective mask for this stage.

The drain-hole at the bottom of the kiln can be opened once the fire is completely quenched to allow the charcoal to drain. The water draining off has a very high pH (alkaline) from the ash produced during the burning stage, and should be disposed of carefully as it may be too strong for plants.

The finished product may have just a few brown ends to remove before the kiln is tipped up on its side to shovel out into sacks.

The main disadvantage of drenching the Kon-tiki kiln is that the charcoal is wet. This is not a problem if the biochar is used on a small scale to make soil improver for personal use in the garden, but if the charcoal is to be transported any distance to be sold, drying it out will be preferable.

Dousing the kiln with water.

The biochar with just a couple of brown ends to be removed.

Ed Armstrong and Robert Taylor bag the biochar.

Making a Version of the Kon-tiki Kiln

The Ithaka institute, who designed the Kon-tiki kiln, is keen to make this an open-access technology; the plans can be requested from them (for contact details, see useful addresses at the end of the book).

BRICK KILNS

Brick kilns are also heated internally (as opposed to the retort, which is heated externally), and run as batch burns. They are perhaps similar to metal kilns but give better yields, mainly

A German brick kiln reinforced with an external steel structure.
MARTIN HALES

Concrete beehive kilns in a battery photographed in Germany.
MARTIN HALES

Operation of a Missouri Kiln of 235cu yd (180cu m) Capacity

Loading	Two days, two people plus loading machinery
Burning	Six days, two people on twelve-hour shifts, or three on eight-hour shifts
Cooling	Twenty days (minimum), one person part-time supervision
Unloading	Two days, two people plus loading machinery

From: *Charcoal Production – A Handbook* (Natural Resources Institute)

because of the better thermal efficiency and re-duced heat loss offered by the brick. Brick kilns are fairly low technology, can be built *in situ*, and have a life span of ten or more years.

Many brick or masonry kilns are built in a 'beehive' or domed construction, such as the Brazilian beehive kiln. A standard kiln of this type would be 11ft (5m) in diameter, with a volume of 64cu yd (49cu m), and would have six out-let ports with chimneys and eighteen air inlet ports. There are also fifty brick-sized emergency outlet ports to prevent a build-up of gases that might cause an explosion, with devastating effect on the kiln. Although individually fairly modest in proportion, the kilns are built in batteries of thirty or forty kilns in one location. Most of the 5,000,000 tonnes of charcoal produced in Brazil is made in these beehive kilns.

The Missouri Kiln

The Missouri kiln, frequently found in both America and Africa, is a concrete and steel kiln rectangle with a domed roof. The capacity is much larger than the beehive kiln, and it is designed to accommodate a front-end loader driven into the kiln for filling and emptying. It is well insulated, and can be operated in cold temperatures without losing efficiency.

Although the Missouri kiln is widely used in America and elsewhere, there has been concern about the emissions from these large-scale char-coal plants. Modifications allow the burning of the gases produced to reduce the release of car-bon dioxide and methane into the atmosphere.

THE FUTURE

Making charcoal in a traditional pit, earth, barrel or metal ring kiln without any attempt to utilize or recycle the gases produced is un-doubtedly now old, inefficient technology. With concern growing over the inevitability and scale of climate change, charcoal burners will have to change their ways. Open kilns, which do not restrict oxygen, are seen to burn hotter and cleaner. However, for large-scale production the future will be retort kilns, described and discussed in the next chapter.

Chapter 5
Retorts

The difference between a retort and a kiln is the heat source. Whereas in the kiln the heat source is internal, in the retort the heat source is external and separate from the timber being converted to charcoal. As the charcoal wood gets hot, volatiles are produced and these gases are diverted back to the external firebox and burned to further heat the wood. This creates efficiencies leading to a higher conversion rate of weight of wood to charcoal produced, and fewer pollutants emitted into the atmosphere.

HISTORY

The history of the retort is bound up with the history of gunpowder, as described in Chapter 1. The first recorded gunpowder mill in Britain was set up in 1530, in Bermondsey, London. In 1543 another mill was recorded in Rotherhithe, also on the Thames. By 1550, a third was started at Chart Mill, in Faversham, Kent. By the 1560s, the Evelyn family (ancestors of diarist and forester John Evelyn) were engaged in gunpowder manufacture in the New Forest, from which they gained considerable wealth. In 1690, Huguenot refugees were employed to build a second mill at the Faversham site, Oare Mill, as the French system of gunpowder production was seen as superior to the British.

Early Metal Retorts

The first documented use of a retort, using an external heat source and a cast-iron cylinder complete with pipes for collecting the distillates from converting timber to charcoal, was in 1798 at the Oare Mill gunpowder works in Faversham. In 1759 the British government had purchased

this site, and a second mill was built nearby in 1787. That same year the government consulted Richard Watson, Bishop of Llandaff, on how to make improvements to gunpowder manufacture in Britain. He suggested the use of cast-iron cylinders for charcoal production. This new technology was soon extended to other mills, such as the Royal Gunpowder Mill at Waltham Abbey, Essex, where charcoal retorts were installed in 1794. Elsewhere, gunpowder works were established in the North West with mills at Sedgwick (1764), Backbarrow (1798) and Elterwater (1824), all using the same technology.

Charcoal Retorts for Gunpowder Manufacture

The use of cylinders for producing charcoal meant that the product was very pure and un-

An old cylinder from the gunpowder works at Sedgwick, used as a gatepost.

the expansion of the British Empire. As part of this demand there was also a brisk trade in another essential ingredient of gunpowder, saltpetre (potassium nitrate), supplied by the East India Company, and obtained from Bengal, India, a resource held by the British since the late eighteenth century.

The industrial revolution was accelerated with the development of charcoal-based explosives, used in mining and for the clearing of routes for roads and rail.

Many of the gunpowder mills survived until the outbreak of World War II, when safety concerns about the mills, often located in highly populated areas targeted by bombing, led to gunpowder manufacture being moved to Scotland and a plant at Ardeer, Aberdeen. This plant remained open until the 1970s, though by then charcoal was being imported from overseas.

Batch Retorts

The potential for exploiting the distillates of wood was recognized throughout Europe. In 1807, French chemist Jean Baptiste Mollerat and his brother started to develop a system for distilling wood tars, which involved creating charcoal in a retort made of cast-iron cylinders housed in a brick chamber. In 1819, German chemist Reinhold Freiherr von Reichenbach developed a similar system, and in 1835 published a paper on the components of wood tar.

In 1850 a retort was developed where sealed cylinders of wood were lowered into a brick

contaminated by the stone and grit that was often found in charcoal produced in an earth clamp. This was extremely important where there was a risk of uncontrolled explosion from rogue sparks. A cylinder had the added advantage that distillates such as acetic acid could be collected and traded to improve efficiency.

By the early nineteenth century, gunpowder production had greatly increased, largely in response to the British government's military demands, not least to fuel an arms race which became crucial in the Napoleonic wars and with

Richard Watson, Bishop of Llandaff (1737–1816

Born in Heversham, Westmorland and educated at Heversham Grammar School and Trinity College, Cambridge, Richard Watson became a professor of chemistry in 1764. He married Dorothy Wilson, daughter of Edward Wilson of Dallam Tower in 1773. In 1782 he was appointed Bishop of Llandaff, where he served for the rest of his life. His interest in chemical processes led to the publication of his paper 'On Pit-coal' in 1781, detailing a system for condensing the volatile products from coke ovens.

Through this and other works, Watson was regarded as an expert on charcoal production and new technologies to capture distillates. In 1788 he bought the Calgarth estate at Troutbeck, Windermere. He is buried at St Martin's Church, Bowness on Windermere.

Bishop of Llandaff's grave in St Martin's Church, Bowness on Windermere.

fire-box chamber, and when the distillation was complete the cylinder could be lifted out and replaced with a new cylinder using the residual heat held in the brick kiln. This form of retort was exported to New York, USA, and similar systems are still in use in Asia. At a similar time Hessel developed an iron retort of 24sq yd (20sq m) capacity, which was the prototype for retorts for another thirty years. Towards the end of the nineteenth century, horizontal tunnel retorts were developed by Meyer where the wood was moved in and out of the kiln on wheeled cradles. Similarly, the Blair kiln also used a railway system to move the timber in and out of the tunnel. This system was seen in use in the Forest of Dean in the 1970s.

The push for efficiencies reached a critical point in the 1920s and 1930s, by which time retorts were being built as a series of chambers or 'batteries' (connected cells), where the heat could be passed from one chamber to another, allowing almost continuous production. These 'batch' retorts were often installed in groups of five, as this enabled gases to be directed to the next one along in sequence. With five retorts, hot gases were always available for wood drying and beginning the process of carbonization. (Ref.:Kelley pers. comm.)

Continuous Burns

A continuous burn, always the ultimate goal of the charcoal retort, was achieved in two ways: with a horizontal retort, and with a vertical retort.

Horizontal Retorts

A horizontal retort was developed to carbonize residues such as sawdust, nutshells and coffee-bean waste. For example, the Aldred Continuous Carbonising Unit is used in Kenya and Tanzania to process coffee husks and cashew-nut shells. These retorts are expensive to install but are very efficient and produce high quality charcoal. The raw material is fed into a hopper and drawn mechanically through the combustion chamber. The gases are taken off to fuel the furnace, which in turn heats a 'jacket'-type chamber around the combustion chamber. Surplus heat is used to dry the feedstock prior to burning.

Another example is a rotating cylindrical kiln, which can be made on a scale to produce fifty tonnes of charcoal a day. These kilns process particulate residues that are suitable for making briquettes. An example of a rotating cylindrical kiln can be found in Sri Lanka, which is used to

The Aldred Continuous Carbonising Unit. DON KELLEY

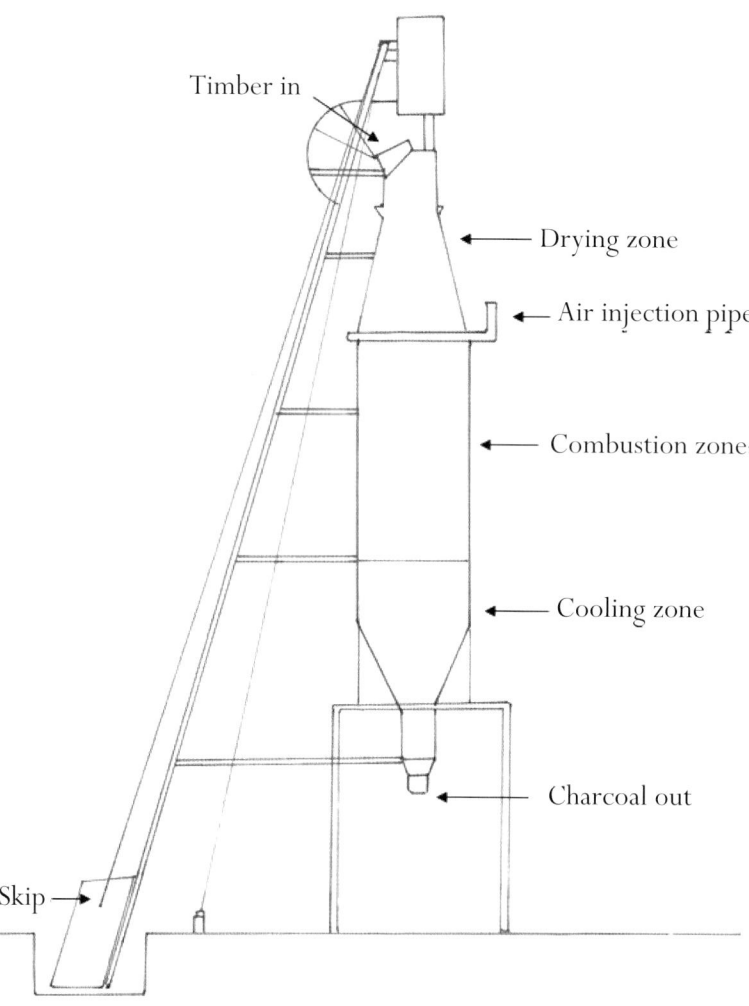

Timber in

Drying zone

Air injection pipe

Combustion zone

Cooling zone

Charcoal out

Skip

Diagram of a vertical retort.

make coconut-shell 'activated' charcoal, which in turn is used widely for industrial, medical and food purposes.

It is possible to purchase one of these retorts: a Chinese version is available from the Wanqi Mechanical Equipment Co Ltd in Zengzhou City. They are sold in 'sets' for $5,000; each set has a 6.5cu yd (5cu m) capacity capable of producing 1,100–1,550lb (500–700kg) per hour, and can operate continuously around the clock. They claim to be able to supply thirty sets a month, so production could be scaled up as required.

Vertical Retorts

Whereas horizontal retorts have been developed for small particles and residues, for processing timber into charcoal, vertical retorts are the more common system. The Lambiotte SIFIC/CISR retort developed by the Belgian company Lambiotte & Cie is currently manufactured and distributed by 'Balt Carbon'. The company is developing retorts in Latvia, which they claim can produce up to 8,000 tonnes of charcoal a year, and as a by-product generate up to 1,000kw of electricity from the excess heat. The timber is raised by pulley to the top of a vertical structure in skips, and released down through the retort. Once processed, the charcoal emerges at the bottom, and the various by-products are also collected. The gases produced are fed back into the system as fuel, and excess heat is captured in hot-water tanks.

Recent Developments

By the 1990s, charcoal production in Britain had shifted into small-scale enterprises with individual burners making charcoal, mostly using metal ring kilns. However, some entrepreneurial individuals wanted to develop small-scale retorts, in which they could harness the benefits of retort technology: better conversion rates and reduced emissions. One innovator, the late Robert Webster, developed a twin-chamber retort that was produced in Kent over a number of years, championed by Carl Cooper of Traditional Charcoal from Cheshire.

THE CURRENT PICTURE IN BRITAIN

Four Seasons Fuels are producing retorts for the domestic and overseas markets, and have recently reported a sales increase year on year. Originally they took Robert Webster's retort design, and after his death further developed it to make a robust static retort, now manufactured by the Birmingham-based fabricators, Pressvess.

The Pressvess Standard Retort

The Pressvess standard retort is a double-chamber kiln capable of producing 770–880lb (350–400kg) of charcoal in one burn, depending on wood species and seasoning. The static batch re-tort, not designed to be portable, is constructed of 10mm steel, which can have the drawback of a tendency to overheat in spite of the inner glass-fibre insulation.

The double boiler-plate charge cylinders are 1.6m diameter. External doors are made with 500mm of ceramic blanket sandwiches in two plates of 4mm steel to help heat retention. The entire inner is lined with 500mm ceramic blanket to minimize heat loss around the cyl-

A Pressvess retort. NICK HARRIS

inders. The company offers to customize the design and size depending on a charcoal burner's requirements. The main difference between the Pressvess twin-chamber retort and the similar Oxford retort (described below) is that the chambers are fired simultaneously, with no drying time built in. To reduce the burn time, Pressvess suggest that a pre-burn is conducted to raise the temperature to 200 degrees centigrade. This would allow the steam to vent out before cooling and relighting the next day –halving the burn time from twelve to around six hours.

One leading exponent of the Pressvess system is Mark Parr, who set up the London Log Company, supplying the London restaurant trade with single-species charcoal. Some of this charcoal is made on the Mereworth Estate in Kent, owned by Lord Falmouth, who is keen to support British business. The London Log Company charcoal is made in two twin-chamber Pressvess retorts, which are kept running day and night, producing about 3.5 tonnes of charcoal a week all year round. Mark notes that the retorts, which take a lot of wear when run continuously, need regular servicing to function efficiently. With the twin-chamber retorts though, the key is to always keep them hot, so as to avoid the contraction of the metal when cooling, which can cause buckling and distortion. Mark points out that although the retorts need replacing fairly regularly, the payback time is acceptable and the profits offered are 'very good'.

He has now expanded the company to employ ten people in Britain and has set up a branch in Spain for the production of high-quality holme oak (*Quercus ilex*) charcoal. In Spain, they use the Ukrainian CK-2 'Euro' retort, manufactured and marketed by Green Power. These retorts based on the twin chamber retort shown in simplified form in figure 101 and have a higher capacity and greater resilience than the retorts in their British operation, producing about 300 tonnes of charcoal a year.

Another Pressvess exponent is Matt Edmonds, who recently moved into charcoal production as an adjunct to his firewood business. Hoping to keep the charcoal side of the business relatively small, he opted for a single-chamber Pressvess retort without the ex-

> ### Cahoon Nuts
>
> Cahoon nuts were imported from Belize during World War I to be made into charcoal for fuel. Coal had been requisitioned for the war effort and domestic fuel was in short supply. Cahoon nuts produce oil and the flesh can be eaten, but the husks are ideal for charcoal, keeping their structure which is shaped like mini-briquettes. 2.2lb (1kg) of charcoaled cahoon nuts gives six hours of heat. Nick Harris of Four Seasons Fuels has connections in Belize, where there is a plan to install ten Pressvess type retorts amongst several Belize villages, in order that the husks can be marketed globally as barbecue fuel, thereby giving a boost to the local economy. The main restriction currently on the success of the British/Belize cahoon nut venture is the high import duty, which makes importation of the charcoal expensive.

tra cooling wagons. He operates it as a batch retort: filling, firing, then cooling, before unloading and starting again from scratch. He is mainly using seasoned wood, with a yield of 200–220lb (90–100kg) per burn.

Almost Continuous Retorts

Twin-chamber retorts have been developed which are 'almost' continuous. This type of retort has the advantage of being able to utilize excess heat to reduce the moisture content in the wood. This allows steam to vent out into the atmosphere and not to circulate in the pyrolysis chamber, potentially cooling the gases and slowing the burn. The timber is loaded into 'wood cradles' with vents at the base to allow first the steam to vent out, and then the heat to circulate right through the wood. At the end of the process, the cradles are removed and the vents sealed, otherwise the sudden inflow of oxygen could cause the wood to ignite and burn away to ash. Thus sealed, the cradles lose heat naturally and when cool enough are opened to empty out the charcoal.

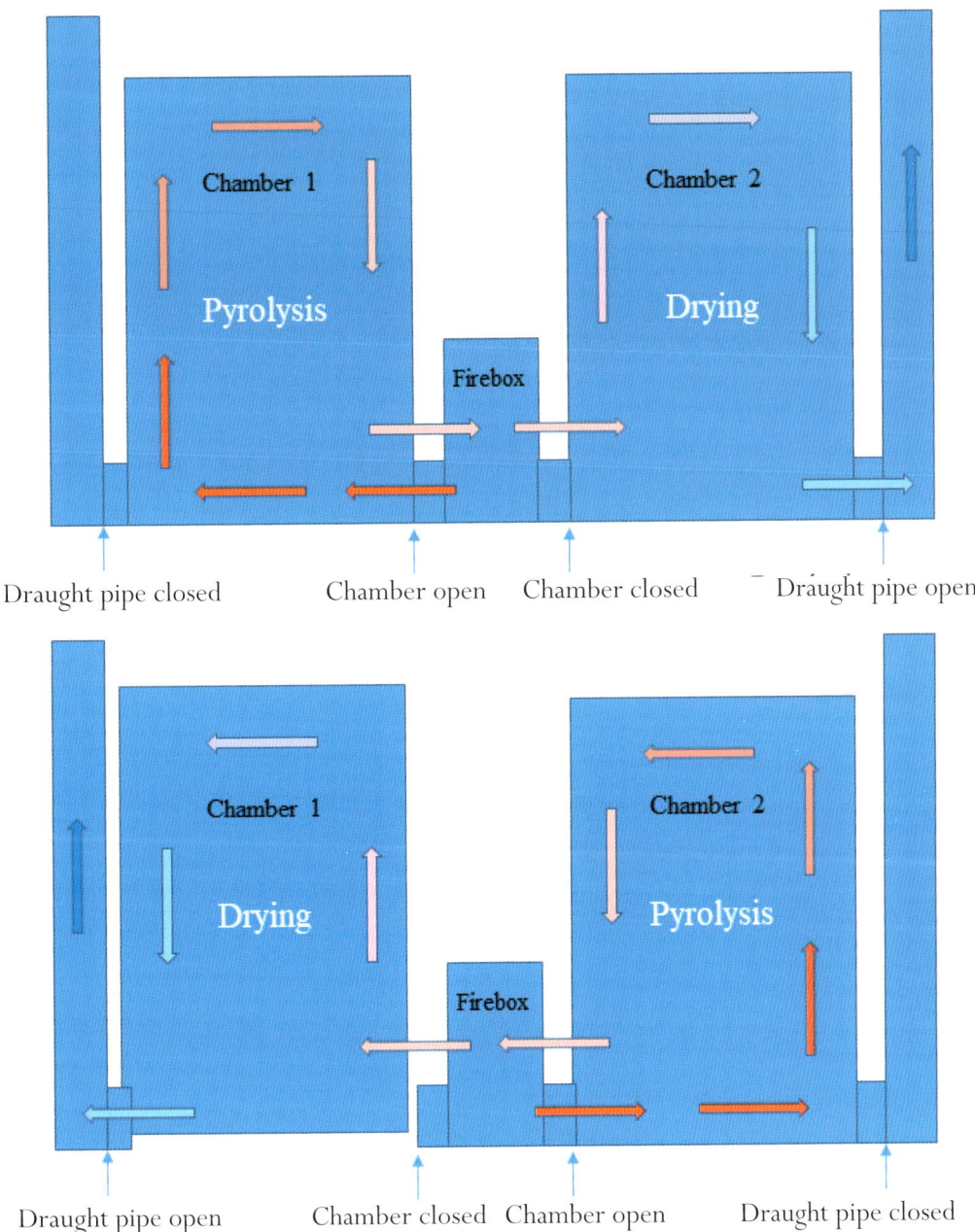

Diagram of a twin chamber retort.

The firebox of the Oxford retort. JOHN PERLMUTTER

The Oxford Retort

The Oxford retort was developed in the early 2010s as a static retort, capable of making one tonne of charcoal a day using a hybrid of continuous and batch technology. Trolleys packed with 2 tonnes of timber are wheeled on a rail system into a metal chamber, which is heated by a flue through the floor. The flue comes from a furnace stoked with waste wood to get the timber up to 280°C, when it starts to produce its own wood gas to burn to raise the heat still further.

The first chamber runs at temperatures of up to 400°C. Any excess heat is passed back to the second chamber where the next trolleys of timber are pre-dried in preparation. When the first chamber has completed charcoal conversion, the trolleys are removed for cooling and more trolleys of chopped and stacked logs are ready to go back in for drying. The valves can then be swapped over, so that the heat from the firebox is diverted into the second chamber which contains the dry wood. Once up to temperature, wood gas is produced, which fuels the burn.

As with most retorts, more gas is produced than is needed to complete the burn. Therefore an excess of heat is created, which has the potential to generate 25kw of energy. The latest version of the Oxford retort has the ability to harness this energy. Although the precise temperature of the optimum burn in the Oxford retort has not been stated, it has been fine-tuned to an impressive degree in order to convert all the cellulose in the wood to carbon, but to leave the lignin (found in the annual rings in hardwoods), which retains the optimum amount of volatiles to impart their distinct flavour to the cooking experience. Single-species charcoal is a speciality which sells at a premium to barbecue cooking connoisseurs.

Static retorts have a high initial investment cost, and issues around the movement and storage of vast quantities of raw materials. The principle of taking charcoal production to the timber resource has not been lost. Interest is growing in the area of small-scale portable retorts, which can be moved from wood to wood. The next section considers some of the technology available for the small-scale charcoal producer.

Loading the Oxford retort. JOHN PERLMUTTER

Timber drying to go in the Oxford retort. JOHN PERLMUTTER

Matt Williams

Matt, a thatcher for twenty-five years, was featured with Monty Don on the television programme *Master Craftsmen*, teaching a group of hopefuls the skill of thatching. However, he became disillusioned with many thatchers' lack of care to do a 'proper job', some of whom appeared happy to do a rush job to make it pay. Matt moved into woodland crafts, setting up business in a wood owned by Oxford University, previously used as a research site. The wood was a derelict coppice of hazel, field maple and ash, which had not been worked since the 1930s. Restoring the wood required working out how to create a market for fairly large timber.

Mark decided to buy a metal ring kiln and start making charcoal. He concentrated on single-species charcoal, and found a ready market amongst chefs keen to use wood-fired cooking, which by 2012 had become a burgeoning trend. However, he was struck by the inefficiency of ring kilns with their low conversion rate of 1:7–1:8. He discovered by accident, as he watched flames shoot 10ft (3m) into the air from the chimneys, that the gases emitted during the burn naturally caught fire. This led him to investigate the availability of retorts, which would utilize the wasted energy and improve the efficiency of the burn.

He was concerned from a business viewpoint that the small-scale retorts currently available would be uneconomic, and would fail to produce enough charcoal to provide a realistic payback on the investment. He initially purchased a retort from Ukraine, which promised well but was destroyed in a freak lightning strike, three weeks after putting it to work. He then had a stroke of luck when he met Simon Fineman of Timbmet (a leading timber company), and he was able to develop his own design retort, with Simon's generous financial backing. The retort runs twenty-four hours a day, seven days a week, and employs sixteen people. He plans to

Matt Williams of the Oxford charcoal company.
JOHN PERLMUTTER

bring a new retort into production, which can produce 2 tonnes of charcoal a day, for which he is confident he has a ready market.

His technology is being adopted in Cuba, which has 50 million hectares of derelict sugar plantations, which became obsolete with the demise of Soviet Russia and the subsequent collapse of the Cuban sugar industry. One third of that land is, at the time of writing, being reclaimed for agriculture, with the clearance of scrub providing the fuel for a budding charcoal industry.

SMALL-SCALE PORTABLE RETORTS

The Exeter Retort

The Exeter retort was developed by Robin Rawle and Geoff Self of the Carbon Compost Company, in response to a demand for locally produced charcoal fines for biochar. It has also proved popular, though, for barbecue charcoal production on a small to medium scale. The retort has a capacity of 2.2cu yd (1.7cu m), equating to approximately 0.75 tonnes of air-dried timber, and can produce 330–420lb (150–190kg) of charcoal per burn, representing a conversion ratio of 1:4. The record is held by Alan Waters, who has championed this retort, with a yield of 436lb (198kg). This retort can be operated by one person and charged daily. It is designed to sit on a trailer, which can be towed to any site with reasonable vehicular access.

Robin Rawle with a new Exeter retort.

Iain Loasby demonstrating how to fill the retort from just one end.

Loading the Kiln

When loading the kiln a handy stack of timber is required, cut to 22in (55cm) in order to pack the logs four rows deep in the inner chamber. It is not essential to cut to 22in (55cm), as it can cope with a range of sizes. The principle is, that the more wood you can fit in, the more charcoal you will get out. Likewise one can burn a mix of species or single species to achieve consistent charcoal.

As there is a door at each end, loading the inner chamber can be at just one end (pictured), which involves getting inside to make sure the wood is packed tight, or from both ends. Using smaller pieces on the bottom reduces the occurrence of brown ends, so it is worth spending time arranging the timber with minimal gaps between, as with building a drystone wall. It is important to avoid using sawn square timber and packing it without any air gaps, which re-stricts the permeation of heat inside the retort, resulting in an incomplete burn.

Once completely full, the retort is ready to be sealed. The seal on the doors often needs to be improved, as the high temperatures inevitably distort the metal – running repairs are part and parcel of using this kind of retort.

The inner door is sealed with bolts, a process that can be speeded up if a cordless drill with a bolt attachment is used.

The firebox on this retort is quite vulnerable to overheating, especially where the gas comes out of the inner flue to be burnt. The latest version of the Exeter has now resolved this issue with a stainless-steel outer casing, which is much more resilient.

The bottom image on page 92 shows the detail of the inner flues, which have a patent on them in the UK, Europe, Australia and America.

Seal the outer doors at each end, if loading from both ends.

RIGHT: *The loaded retort.*

BELOW: *Iain uses a mastic gun loaded with fire cement to glue a new section of ceramic lagging rope to the door.*

Doing up the bolts with a cordless drill.

Detail of the inner flue, which brings the gases down to the firebox (patented).

The outer door also needs sealing, and the bolts tightened.

Lighting the Kiln

The fire can be set going in the firebox using waste wood. Softwood such as pallet wood is ideal, and you would use about four pallets' worth of wood to get the retort up to temperature. Although any preferred method of lighting is entirely up to the burner, paper and kindling would be the minimum requirement. It should be sufficient to light from one end, but position the retort to get a good airflow into the firebox to help get it going. Continue to feed wood in through the firebox door, and keep an eye on the temperature gauge. One burner provides a community service – 'shed removal', taking down and disposing of old sheds and fences to fuel the furnace of his Exeter retort.

The thermometer is connected by sensor to the vent where the gases leave the charcoal chamber. For this first stage the vents are open to allow water vapour and early volatiles to vent out to the atmosphere. It can take two hours to get up to 400°C.

The caps are then put on the vents, first sealing the one furthest from the direction of the

The firebox with a fire of waste wood to build up heat.

The internal temperature of the retort is monitored on the outside.

wind. The temperature usually drops to around 350°C at this point, as there is still water vapour present. When directed down the internal flues to the firebox, this steam can quench the fire, but soon the volatile gases start burning and the temperature begins to rise again. The maximum desired temperature is 500°C, and it is best if the retort is controlled so that this is achieved slowly and steadily.

The Second Phase – Burning the Gases

Once the retort has reached the target temperature and the vents are sealed (allowing the gases to burn in the firebox), the firebox doors can be closed. As this is done, a roaring sound should be heard as the gases ignite. As long as the temperature does not drop too much, the process can mainly look after itself. It is important, though, to be present to tweak valves and vents to keep the temperature steady. However, this process can be eased a little with the use of an automated vent, which, when added to the retort, regulates the temperature by releasing excess gases into the atmosphere.

Vents open to release steam.

The retorts now come with a sensor to open and close vents automatically.

The finished retort; note that the volume will reduce by 50 per cent.

There will always be more potential heat produced than is required to complete the process. The excess is used in larger or more complex retorts to pre-dry timber. The gas burning stage can last for six to eight hours before the volatiles are spent and the retort temperature drops down below 350°C: time happily spent on other tasks.

Cooling

After at least six hours the retort temperature should drop below 350°C, the sign that the volatiles have gone and that nothing is left of the fuel in the firebox. All that remains is to wait for the charcoal to cool, when it is simplest to do nothing, in the age-old, traditional way, and leave the kiln overnight. It may be necessary, however, to wait and watch the gauge reading until it drops to 40–50°C before opening the retort. The danger, as ever, is re-ignition, because even if charcoal has not been exposed to flame, the combination of heat and oxygen can cause spontaneous combustion. A handy water

supply should be always be available at this stage in case of re-ignition.

The structure of the charcoal produced in a retort is very much intact, but it is also quite friable, so even the biggest lumps of charcoal will break down into manageable pieces as it is put into sacks.

The final weight of charcoal taken out of the retort will depend on the wood species used and the efficiency of the packing of the timber into the chamber. For example, the yield for the burn in the retort pictured was 330lb (150kg) – although, according to the company which makes this retort, the maximum output is 420lb (190kg). Another user of the Exeter retort whom we spoke to was happy to achieve 290lb (130kg) with sycamore. Properly pre-dried, dense timber such as oak or beech that is well stacked, will maximize yield in any kiln.

Carbon Gold

Carbon Gold, a leading biochar production company (*see* Chapter 9), has produced two re-

Syngas

Syngas is an abbreviation for synthesis gas, a mixture of carbon monoxide, carbon dioxide and hydrogen. The syngas is produced by gasification of a carbon-containing fuel to a gaseous product, which has some heating value. Syngas production includes gasification of coal emissions, waste emissions to energy gasification, and steam reforming of coke.

From: http://biofuel.org.uk/what-is-syngas.html

torts specifically for making biochar. The first is the SuperChar 100 Mark II, with a capacity of 2cu yd (1.5cu m), and the second is the Super-Char 500, with a capacity of 15.7cu yd (12cu m) over two chambers. These retorts will burn any woody feedstock from rice husks to 12in (30cm) length logs. The initial heat source is from burning dry woodchip in the firebox until the internal temperature is high enough to release the wood gas or 'syngas'.

The main advance of these retorts over others discussed so far is that they have a system for blowing the heated air back into the furnace after it has circulated around the kiln body. This allows feedstock with a higher moisture content to be charred without reducing the working temperature of 450°C.

The Carbon Gold SuperChar 500 retort. SIMON MANLEY

THE HOOKWAY CHARCOAL RETORT

The Hookway retort was developed by James Hookway and is based on the idea of a 'Kelly Kettle' or rocket stove. The rocket stove is designed to boil water efficiently when poured into a metal 'jacket' surrounding a central flue, where a fire is set and fed from the top through the chimney into the rocket stove section. The effect is a strong updraft of air, which fans the

The wood is packed round the central flue.

flames to get a rapid rise in temperature, which boils the water.

The Hookway retort employs the same principle of a jacket and flue as the Kelly Kettle, packing wood for carbonization into the jacket around the flue.

The firebox is in the centre at the base of the retort cylinder. When lit, a good draw of oxygen creates a syphon effect as the gases burn up the chimney, heating the wood in the surrounding 'jacket'.

To use the Hookway, light some kindling in the firebox at the base of the retort, and feed with dry kindling or relatively small wood as the fire takes off.

When the kiln is up to temperature the gases produced are fed down to the firebox at the base, and ignited to keep the burn going. It is important to keep an eye on it, because at this stage if the temperature drops, more sacrificial wood may be needed to bring the retort back up to the temperature required to produce wood gas again. Much of this monitoring is done by listening for the 'roar' of the burning wood gas.

After about four hours the gases will be spent and the retort can be left to cool. As with all charcoal making, ensure that the temperature has dropped sufficiently to avoid the danger of re-ignition when oxygen is introduced. As al-

The Hookway rocket retort.

AlanWaters lights the Hookway retort, watched by James Hookway.

The retort quickly gets up to temperature and the wood starts gassing.

The charcoal is high quality, as in a larger retort.

James Hookway bagging the charcoal with the retort tipped on an angle.

ways, keep a water supply to hand as a standard safety provision.

The charcoal produced from a Hookway retort is good quality and will yield about 55lb (25kg). To bag the charcoal it is easiest to tip the retort on its side. One option is to strap it to a trolley, which makes it easier to move around.

The wonderful thing about James's invention is that he has set it up to be possible for anyone to make their own version, as he has drawn up construction plans which can be purchased via his website. Details are in the Useful Addresses section at the back of the book (*see* page 173).

CONCLUSION

In a world where resources are truly finite and population growth outstrips our ability to innovate our way out of danger, it is essential that we embrace technology with 100 per cent efficiency saving, offering the opportunity to collect and utilize waste products, rather than release them back into the atmosphere. The options currently available to the small-scale burner are either too small to be economic or too expensive to fit a realistic business model. 40 per cent grants are currently available for forestry businesses to purchase equipment, making the initial outlay costs look more realistic. While building your own is also an option, the hours given to developing and perfecting these retorts means the time taken to build your own at living-wage level, makes the cost of a ready-made retort seem not so very expensive. *See* the Appendix for a comparative chart of all forms of charcoal-making equipment.

Chapter 6
Bagging and Marketing Charcoal

Once you have made charcoal, you may find that you become smitten by the process and will soon have made far more than you and your friends and neighbours can consume – so now is the time to find a market and go into serious production. This chapter first considers the marketing of barbecue charcoal, and then the similar but different issues when selling biochar.

As artisan charcoal burners we can be justly proud of the quality of the product we produce. If this were more widely understood and appreciated, it would probably be much easier for burners to sell their wares. However, the key issue for British barbecue charcoal is competition with imported charcoal, which is stacked high and sold cheap, often in brightly coloured bags

Helen and Sam examine the contents of the kiln shown burning before on page 65.

Charcoal bagged and ready to go straight to the customer. DUNCAN GOULDER

or labour-saving, pre-packed, easy-light barbecue trays. A less exciting bag of charcoal without 'come and buy me' packaging may be left on the shelf, particularly if it is twice the price. There are signs that the slow food, buy local, eco message is having some impact on buying choices, but it is unlikely that this will be uppermost in the mind of a busy person who stops off at the garage on a Friday evening to buy charcoal for the weekend barbecue. Focusing on the 'high end' gourmet market may be a good strategy, as it avoids the issues around price sensitivity. One thing is certain: you will not be competitive on price with cheaper imports, so you must find a way to persuade people to buy your charcoal.

As with British barbecue charcoal, the market for biochar is very select, despite our knowledge of biochar's positive impact as a soil improver – a message that has not yet permeated the general public's consciousness. The small but steady market for peat-free compost – an alternative to the many millions of tonnes of peat-based composts sold in Britain and the world – challenges a trade which is unsustainable and damaging to the environment. Although the issue is periodically brought to the public's attention, for too many the price barrier ensures that this trade persists. Biochar must surely have a role to play in the development of alternative growing media, and educating the public is an ongoing task. The biochar debate is having an impact, raising awareness of soil health issues, but there is much more work to do to promote the use of biochar as a beneficial soil improver.

Iain Loasby modelling the Powercap Active dust mask.

Twiggy with a cartridge-style dust mask.

It is therefore important to get the 'look' and the message right for all forms of charcoal: from the bags and tubs chosen in which to sell the products, to the accompanying promotional materials.

EMPTYING YOUR KILN OR RETORT

Charcoal cannot be graded and bagged without getting very dirty – but the good thing is, it is pure carbon dirt! It is important at this point to raise some health and safety issues associated with charcoal dust. Unlike smoke, which contains a range of toxic volatiles, charcoal dust is not considered actively toxic, but is classified as a 'nuisance dust', which can irritate and cause congestion in the upper and lower respiratory tract. It is neither advisable nor pleasant to breathe in charcoal dust, so a dust mask is essential. The one modelled by Iain on the opposite page is a Powercap Active, one of the more deluxe models available, apparently great for beard (and glasses) wearers, as the motorized

pump keeps a good seal around the face – and this mask eliminates the trademark eye make-up and blackened face which are the hallmarks of using the more basic dust-masks sold in hardware stores (which if nothing else prove you have been working hard).

A range of different quality dust masks is available, of course, but the simple paper versions only cover the mouth and nose, and do not offer a good level of protection, so should only be used if bagging charcoal very infrequently. For those actively working with charcoal, a cartridge type should be sufficient, but it is important to remember to change the filter regularly.

Bagging Straight from the Kiln or into Bulk Bags?

For a small-scale producer, when refilling the kiln can be delayed, the best practice is to grade (sieve) the charcoal straight from the kiln into bags, which reduces the need for double handling. Charcoal is quite a delicate product and will start to crumble if handled too roughly or too often. One consideration for the small-scale

Grading and bagging straight out of the Exeter retort.

Weighing the bulk bag to gauge the yield of the retort.

producer, especially if working alone, is to be able to draft in help at the grading and bagging stage, as it is quite difficult to unload a ring kiln efficiently alone. For lone burners it is virtually impossible to grade and bag in one operation. The retort pictured below lends itself to lone working, which may well be a consideration if you are determined to keep it simple and do it yourself.

A word of caution: it is important to ensure that the charcoal has been sufficiently 'cured', or exposed to the air for long enough to ensure there is absolutely no danger of re-ignition. Stacking the bagged charcoal and leaving it for twenty-four hours before delivery is best practice, to allay any concerns the customer may have about re-ignition. This is a problem when

faced with a sudden heat-wave and a spike in demand, which means there may be a temptation to get the charcoal delivered as soon as possible, when it may not have fully cooled. An alternative approach to bagging direct from the kiln into small bags is to fill bigger sacks or a builders' dumpy bag with the charcoal, and to bag it into smaller bags elsewhere. This has the advantage of being able to bag under cover, and to protect the small bags from rain and dirt.

Charcoal Graders or Riddles

Grading is a term used to describe the sieving of charcoal prior to bagging. A grader can be a simple mesh sieve (*see* below), with a method of catching the 'fines' that fall through. If mak-

A simple grader.

A drum grader that can be motorized or turned by hand.

ing a mesh sieve grader, use 10mm gauge mesh, the minimum size for barbecue charcoal. When processing charcoal for biochar a different approach is used, as described later in the chapter.

Some charcoal burners bag direct from the kiln into bags using a tined fork, which acts in the same way as a riddle, letting the smaller particles drop down to be shovelled up later. Though some of the finer dust tends to find its way into the bags, as long as this is only a small percentage it will not affect the overall quality.

A more sophisticated grader is a drum which can be turned either by a handle or by using a simple twelve-volt motor: a windscreen-wiper motor attached to a car battery works well. The advantage of a motorized grader is that it leave your hands free to bag whilst the drum trundles away tumbling the charcoal.

Brown Ends

Part of the grading process is screening out any brown ends or par char (occasionally called 'brands'), which may have missed being converted to charcoal. There are always some brown ends left in a metal ring kiln, as the process and timing of shutting down is a compromise between stopping the burn too late (which means more of the charcoal may become ash), and stopping the burn too early, only to find that some of the charcoal will have only partially converted and must be removed. Brown ends are not a waste product, however, as they can always go back in the next burn either as sacrificial wood to raise the temperature in the kiln or to complete their conversion to charcoal.

It is easy to tell whether a piece of charcoal is converted completely by trying to break it in one hand. Properly converted charcoal should feel light and will snap readily. If it feels heavy and gives a solid sound when tapped against the kiln, it is a brown end and should be removed. Retort burns are less likely to contain brown ends provided attention has been paid to getting the retort up to temperature, which is maintained until the gases exhaust themselves and the fire goes out of its own accord. Occasionally, in single-barrel retorts loaded with green wood, the moisture in the wood has such a cooling effect that the gases are not completely released, resulting in part-burnt charcoal or brown ends.

BAGGING BARBECUE CHARCOAL

Bags

Paper, plastic, boxes, buckets, foil trays? The choices are limitless. For the new charcoal burner starting out in business, having bags specially printed is quite a commitment, because a minimum print run is likely to be in the thousands. However, cheaper options are available for someone just putting their toe in the water.

The simplest option is to design a label that can be printed quite cheaply, and paste it on to a generic brown bag. This kind of labelling can look good, but if it is not high quality it may appear unprofessional, which could be detrimental to sales. Another option is to purchase generic British Charcoal bags with space for a personalized stamp or label. The National Coppice Federation has updated the bag originally designed for the Coppice Association (which folded in 1996) and made these available to anyone who is a member of a coppice group affiliated to the NCFed. The bags are printed by Selway Packaging (*see* Useful Addresses), who have supplied them continuously since the early 1990s. The brown paper bags have a black line-drawing on the front and information about charcoal burning and coppicing on the back, a model which has become recognizable as a brand over more than twenty years of use (*see* overleaf).

Brown Bags

The benefits of the brown paper bag is that the packaging is relatively carbon efficient and non-polluting in its manufacture and disposal. The down side is that it gets dirty very easily and is soon spoilt by rain, which makes the surface pucker. This kind of packaging seems to appeal to those moved by ecological concerns, but perhaps less so to those attracted to a more colourful modern look.

Glazed Paper Bags

The British hardware store B&Q created charcoal bags which proved a good compromise between ecological concerns and colourful marketing. The bags were made from glazed paper with coloured inks and photographs, and included a strong message linking the product to sustainable woodland management. More importantly, they were both attractive and easy to wipe clean, which was a boon when bagged direct from the kiln or retort.

Plastic Bags

The advantage of plastic bags is that they are waterproof and thus can sit outside in the wet with no risk of the charcoal spoiling, which can be a good selling-point on forecourts and market stalls. The down side is that any residual moisture in the charcoal will sweat inside the bag,

Boxes and Novelty Packaging

Although boxes and novelty packaging may sound like a good idea, the expense must be passed on to the customer. It is important to do careful market research and cost the required business model, to ensure that the charcoal is not priced out of the market because of overly expensive packaging materials. The box pictured (*see* right) proved fiddly to assemble, and confusing because it was not immediately recognizable as charcoal.

A box that can be opened to make a spout for pouring.

making it appear wet, which can be off-putting, although it may not affect the quality. If bagged immediately after it is fully cooled and thus safe from re-ignition, charcoal sealed in plastic bags can offer more resilient packaging. The use of plastic bags is, however, increasingly seen as environmentally problematic.

Weighing or Volume: Skill and Judgment?

The implications of weighing charcoal versus a volume measure, and whether to state the weight and volume on the bag, are discussed in detail in Chapter 7. In this chapter some relevant issues are covered about bagging in relation to selling by weight or volume. Good charcoal is light charcoal, so if selling by weight, the better the quality of the charcoal, the less the economic return in terms of bags sold from any one burn. This is clearly a negative outcome, hence many burners favour bagging via a simple volume measure – for example, a 2gal (9ltr) black bucket is roughly equivalent to 6.6lb (3kg) of charcoal by weight. The beauty of this system is

ABOVE AND FOLLOWING PAGES: *A montage of barbecue charcoal packaging from around the country.* VARIOUS

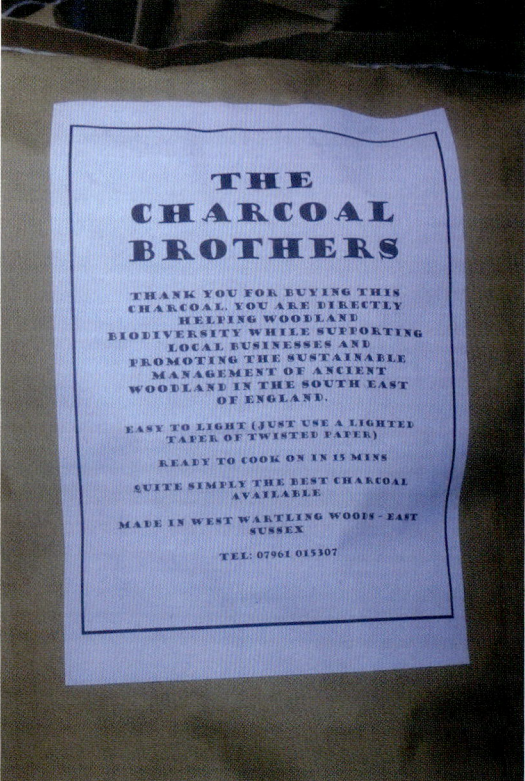

that the bags always look the same in terms of quantity whatever the species of charcoal, and some types weigh more than others.

Whether marked in weight on the bag or not, it is necessary to decide whether to market the charcoal in one-size bags or in a range of sizes. The most widespread size is equivalent to 11–13lb (5–6kg), but a half-size bag is also popular and contains sufficient charcoal for a couple of barbecues. Restaurant-grade charcoal will normally be sold in bigger bags, equivalent to 22–26lb (10–12kg).

Sealing the Bags

How to seal the bags so that charcoal dust does not leak out? The National Coppice Federation bag has one end sewn and the other open at the base for filling, which can then be rolled and stapled shut. Three turns are recommended, and a tough stapler is needed to get through the layers of brown paper. When stapled, the bags sit comfortably on the stapled edge and the stitching on the top looks more professional. The Arrow

A very handy tipping scale.

Ian Taylor using a stapler to seal the bags.

P35 Plier Type Stapler is a good heavy-duty stapler available from the Woodsmith store.

Better still, one can acquire a stitching tool and stitch the open end, but this is yet another outlay to consider. A.T. Sack Fillers (*see* Useful Addresses) sell a hand-held one (at the time of writing priced at approximately £450.)

Sealing plastic or polythene bags is easier, using a heat sealer – a less complex machine costing £50 to £100 pounds.

Bagging Plants

If going into charcoal production on a grander scale it will be necessary to consider a bagging plant. This usually consists of a hopper for the charcoal, which requires a tractor front-end loader to lift a 'dumpy bag' of charcoal into the hopper. There is also a mechanism to weigh the charcoal into the bag, or a measure of volume quantity, and finally another mechanism to sew up the bag.

A potato bagger adapted for charcoal.

In the photograph above a cheaper compromise for the job has been found, using a potato bagger bought at a farm sale for £200. With the addition of an industrial vacuum cleaner to extract the dust, and a modification of the weighing scales to take the lightweight bags, the system works well. It is not at all complicated and can be operated by one person, with the charcoal stored in feed sacks, which can be lifted and tipped into the hopper. An electric supply is needed in order to operate the conveyor belt and extraction process.

MARKETING BARBECUE CHARCOAL

The unique selling points for local charcoal (wherever it is made) are summarized as:

- Low carbon footprint (minimal transportation)
- Supports sustainable woodland management
- Supports local businesses
- High quality production methods

Of these four unique reasons to buy British charcoal, the quality issue tends to hold the most sway with the barbecuing public. There are plenty of jokes at the expense of the Great British Barbecue about smoky fires drenched in lighting fuel, reluctant to burn while the guests remain hungry. As noted above, though, good charcoal if kept dry and unadulterated will be a very different experience. In fact the more likely danger is that the barbecue will be so hot that the food will burn, through overzealous loading in of good charcoal and getting the temperature too high.

But unless people understand and have experienced a good barbecue with local charcoal, they are unlikely to get past the price barrier, to try local charcoal, and become converts because of quality and positive feedback. Building repeat custom with people who experience the benefits of local charcoal can be a long, slow way to grow a business. However, it can be effective on a local level, starting with persuading shops and garages to stock the charcoal.

The practice of centralized buying can be frustrating for the small-scale charcoal producer, as many of the major supermarkets, garages and even village/corner shops are under franchise to a company that will insist that all products are bought from their centralized warehousing system.

Burner Networks

Historically there was the scheme run by the BioRegional Charcoal Company (BRCC), set up by Pooran Desai, who created a network of local burners to supply B&Q stores around the country. BRCC took orders from the centralized B&Q buyers, passing them on to the local burner responsible for the stores in question. The burner delivered the charcoal and billed BRCC for payment, and B&Q were invoiced by BRCC. It worked well for some ten years until a combination of events saw the company withdraw fol-

---

Places that Stock or Buy Barbecue Charcoal

Garages
Butchers
Farmshops
Delicatessens
Garden centres
Caravan and camping sites
Fuel merchants
Corner shops
Grocers
Hardware stores
Farmers markets
Restaurants

---

lowing a series of bad summers, unsteady times for B&Q, and the local supply chain's inability either to expand to meet new markets (BRCC was also involved with Sainsbury's) or to cope when demand spiked at peak season. However, most burners involved were small scale – typically using one to four ring kilns – so it was perhaps inevitable that the business arrangement experienced strain and eventually ended.

Various business plans drawn up over the years for co-operative working or centralized bagging plants have proved difficult to achieve, given the isolated nature of the work and the variable levels of skill and understanding of marketing. The time may well be ripe for a new look at how charcoal burners can co-operate to meet market demands.

Internet Sales

The main difference today from the earlier business ventures of the 1990s is the scope of internet sales. Successful burners speak of the importance of blogs and social media to spread the message and seek out new customers. By far the biggest burners in Britain are supplying 'gourmet charcoal' to the restaurant trade. The premium-priced charcoal is popular with amateur chefs too, and the trend is evidently growing for this type of cooking: definitely a good-news story of effective marketing, conveying the salient points of quality charcoal to the attention of a wider (and wealthier) audience.

A Word on the Weather

The most variable factor with sales of barbecue charcoal is, of course, the weather. A settled period of high pressure and reliable sunshine can see sales soar. A fine Whitsun bank holiday is probably the peak moment for sales at the end of May. However, the season does extend from Easter (which can mean your stockists want deliveries at the end of February or early March) right through to Bonfire Night in November. But the pattern of recent summers in Britain has been very changeable weather through July and August, and this makes it hard for people to plan to have a barbecue party and be confident that it won't be a washout. Stockists become nervous of buying in in any quantity in case it is left on the shelf at the end of the summer. Therefore you really do have to get in early and make the most of a fine April, May and June. This is less relevant if you have tapped into the restaurant trade, and finding all-year-round markets is an astute business model.

GRADING AND BAGGING BIOCHAR

Grading Biochar

Biochar, if processed as a by-product of barbecue charcoal, will have been through a 10mm or 12mm grader already, and it will be acceptable to go from this size particle down to dust.

Bagging straight from a Kon-tiki kiln.

Carbon Gold biochar. SIMON MANLEY

Oxford biochar.

Down-to-Earth biochar from the Isle of Wight. DARREN HOPKINS

Caution is needed though, because too high an ash content can lead to a very high pH (too alkaline). If biochar is the main product, and is being made in a ring kiln or a retort, the charcoal will need crushing or grinding down to this size. Machines are available for this task, but again it will depend on the scale of the operation as to whether purchasing additional plant is cost effective. The author has seen this achieved by means of driving a vehicle over the charcoal lying between two tarpaulins in a yard. If making biochar in a Kon-tiki kiln or similar retort, the particle size is often already fairly small due to the twiggy nature of the feedstock and the high temperatures of the burn.

Curing Biochar

Leaving your biochar to weather outside where the rain can pass through it will leach out any remaining acid or alkaline contaminants (acetic acid from the wood vinegar, or calcium carbonate in the ash).

Packaging Biochar

Biochar packaging issues are both similar and different from those encountered with barbecue charcoal. Keeping the biochar dry is not an issue, as it will be used as a soil improver. The moisture content will, however, have an impact on weight, which will have implications for the cost of carriage if sold online. The most suitable containers for biochar are plastic bags or tubs, which can be purchased plain and a customized sticker then added or printed. Containers can be obtained in a range of sizes, able to hold any weight from 11lb (5kg) up to 44lb (20kg).

If the biochar business is an adjunct to a barbecue charcoal business, the biochar may be sold as a raw product without any additives. Information should be included, for example, about how to use it in the garden to best effect, with advice on avoiding the possible short-term reduction in fertility caused by using too large a quantity of neat biochar on impoverished soils. (*See* Chapter 10 for details on the use of biochar.)

Marketing Biochar

Garden centres are possibly the most obvious outlets, but pricing will be a key issue, as garden centre mark-up tends to be very high, and the bigger commercial garden centres prefer centralized buying. Direct sales of bulk bags of biochar to nurseries and even large gardens, such as those open to the public, might be worth trying if a sympathetic head gardener can be persuaded to trial the product. Promoting the biochar to specialist plant societies may be worthwhile, such as the orchid growers or Water Gardeners International (*see* Useful Addresses). It may also be worth trying to get an article about biochar published in their newsletters.

The internet – blogging and social media – offers potentially useful opportunities to help raise the profile of this relatively new product. Apparently even the shopping channel can be an effective sales point. Chapter 10 discusses further ideas for novel uses for charcoal, and some thoughts about how to get creative with marketing strategies if charcoal or biochar sales are slow.

Chapter 7
Legislation and Regulation

Legislation and regulation tend to be 'dry' subjects, yet as charcoal burners, whether we are running a small or a larger business, we cannot afford to avoid these issues and pretend that the rules do not apply to us. For a potentially hazardous activity, charcoal burning has relatively few regulations. However, for someone setting up as a new charcoal maker, either for barbecue charcoal or biochar, there are some salient health and safety legislation points that require attention: in this chapter we address these issues.

The following are key regulatory questions when setting up a charcoal-burning business and/or a new site:

- Do I need planning permission?
- What may happen if my neighbours complain to the Environment Health Department about smoke nuisance?
- Should I get insurance?

- Should I register under the REACH regulations (Registration, Evaluation, Authorisation and restriction of Chemicals) which control production of chemicals in Europe and the UK? (Charcoal making is regarded as chemical manufacturing)
- What regulations apply when packaging and marketing charcoal?
- Should I consider signing up to the various certification schemes to validate sustainability claims?
- What are the definitions of biochar set down by the leading industry partners, and how can I meet these?

PLANNING PERMISSION

What planning permission, if any, is required when setting up a site to burn charcoal? Planning rules appear to be differently interpreted

A woodland setting for a mobile kiln should not require planning consent.

in different regions, and some of the differences depend on whether the site is urban or rural. The failsafe advice is to consult and check with the local planning department, as it is always good practice to inform the local council of plans, even if it would seem that planning were not needed. A discussion prior to any problems arising can save time and money in the long run.

A mobile kiln in a woodland situation certainly should not need planning permission, as it is a temporary site. The issue of planning arises once a burn site is based in a yard or depot, with static plant. Change of use may need to be obtained if some other activity has been undertaken on the site. If the intention is to stay with the kilns or retort(s) during a burn, then local residents, and the planning office, may be anxious that you are trying to obtain residential rights, albeit 'by the back door'. There is a precedent, however, already set by other charcoal burners, of planning permission granted for temporary residence, on the basis that the kiln(s) need tending around the clock. This is sometimes granted to an individual conducting the burn, or in other cases attached to the activity, which is better still as it allows more flexibility. Planning is granted as long as charcoal is being made on the site, needing someone to stay and tend the burn.

Other Bodies to Inform

If you are just starting charcoal burning, or you are moving to a new site, it is a good idea to make a call to the local fire brigade to inform them of the location and duration of the burn; and arrange to call them when you have finished, too. This ensures that the local fire engine only races to the site if there is a problem and you need help with a genuine fire; so if someone is alarmed by the pall of white smoke (kiln burns) and phones the emergency services, they will at least be informed that it is intentional and not out of control. Do also inform nearby neighbours so that they are warned and do not have a nasty surprise if smoke starts rolling over their garden.

ISSUES OF STATUTORY NUISANCE

Neighbours may become unhappy with smoke from kilns or even retorts. Although the smoke is certainly less from retorts, there is always a tell-tale aroma of woodsmoke when the retort is first lit. Someone concerned about smoke can turn to the local council Environmental Health Department (EHD). The legislation is the Statutory Nuisance: Environmental Protection ACT Pt 111 1990. 'Statutory Nuisance' is defined as 'unlawful interference with a person's use or enjoyment of land or some right over or in connection with it'. The nuisance relevant to charcoal burning is defined under section 79 (a) as follows:

> *Any, dust, steam, smell or other effluvia arising on industrial, trade or business premises and being prejudicial to health or a nuisance.*

Considerations taken into account by EHD inspectors include the following:

- Locality
- Duration
- Sensitivity of the plaintiff
- Intention of the plaintiff
- Utility of the defendant's conduct

If the local authority thinks there is a case for statutory nuisance, it is obliged to serve an 'abatement notice', which may either prohibit the activity, or outline steps that must be taken to modify the effect. This can be appealed against within twenty-one days, but failure to comply is a criminal offence. If an appeal is to be successful it must be proved that the 'best practicable means' were taken to prevent or counteract the effects of the nuisance.

In an experience where neighbours complained of smoke from the author's site, the EHD asked to be informed when a burn was to be lit, in order for their officer to come and take a photograph of the smoke reportedly causing a nuisance. In fact the smoke effects were never seen to recur in the same way as originally reported, and the case was dropped.

The smoke from a kiln can be quite extreme in the early stages before the chimneys go on.

ISSUES OF INSURANCE

It is prudent, as a burner, to get public liability insurance. Woodland owners may already have this, and will only need to inform the insurers that charcoal is now being made. Insurers will usually have queries about 'open fires', but burning in a closed kiln should not be considered an open fire. A barrel or a Kon-tiki kiln may perhaps be considered 'open'.

Where a charcoal burner is not the owner of the wood it is sensible to buy public liability cover, and this would afford basic legal protection if a member of the public were to come along and burn themselves, or trip over your spade.

If planning to have helpers or volunteers under the burner's instruction it is your duty to have employer's liability insurance, to be sure that everyone is covered in the unfortunate event of an accident.

Risk Assessments

It is always wise to do a risk assessment for a burn site, and this will be essential if any public groups, especially school groups, are invited on site to see the operation in action. Risk assessments can seem like a tick-box exercise of stating the obvious, but experience shows that it is good to get into the habit of regularly sitting down and thinking through the possible scenarios that could arise, and the steps to take to avert problems before they occur. It may also be a condition of the public liability insurance cover taken out, or an employer if you are contracted to work off site.

Basic Health and Safety

Basic personal protection equipment (PPE) in the form of dust masks was outlined in Chapter 5, and it is worth reiterating that dust masks are essential with any operation involving charcoal dust to help avoid respiratory problems. Other protection includes taking obvious care to avoid burns and scalds: a moment's inattention can lead to a nasty burn. Wear a good pair of fire-retardant gloves, while strong boots are an essential requirement as the fire or steam can escape at foot level. Managing a charcoal burn, particularly on a warm summer's day, can be extremely hot and sweaty, but do ensure that you are dressed appropriately when coming into contact with fire, heat, steam or hot smoke.

Physical Stresses

Taking care of your body is another essential requirement of the working life of a charcoal

Get appropriate help when lifting heavy items.

burner (and anyone doing woodland work). Years of heavy lifting takes its toll on the joints and back. Doing things by hand without mechanical help can seem attractive, but any labour-saving devices, such as firewood processors and timber grabs, should be considered when planning a long-term future in charcoal making. The benefits of doing warm-up exercise in preparation for hard physical work cannot be over-emphasized.

REACH LEGISLATION

Charcoal is considered a chemical, and all chemicals must be registered under the regulation 'Registration, Evaluation, Authorisation and restriction of Chemicals' (REACH), which came into force on 1 June 2007 and has been brought in over different stages, due to be completed on 1 June 2018. The main enforcement bodies in Britain are the Health and Safety Ex-

The costs of REACH registration

Size in tonnes per year	Registration with EHCA	Letter of Access	Substance sameness test
1–10	£55 approx.	£1,400	Costs shared between registrants
10–100	£55 approx.	£6,500	Costs shared between registrants
100 +	£55 approx.	Not known	Costs shared between registrants

ecutive (H&SE) and the Environment Agency (EA). An attempt was made to exempt charcoal from this legislation, but it failed. Notably, the coke industry was granted exemption for their product (carbonized coal). When registered, information is collected about that substance and shared through the Substance Information Exchange Forum (SIEF). A dossier of data relating to the substance is compiled, and registrants pay for a 'letter of access' to the information. The substance sameness test is a shared cost, dependent on the numbers registering.

Charcoal manufacturers are required to meet the obligations of the REACH legislation, and the charcoal industry is regulated by European law. Despite the UK negotiating to leave the EU, charcoal burning is likely to remain within this legislation for some years. The legislation was instigated initially by Britain, so it would seem that this element of regulation will not be a priority for British legislators to revisit in the short-term, so in all likelihood it will be adopted into British law as it stands.

Prior to REACH legislation, there was a lengthy pre-registration phase where many charcoal makers signed up to register their intention to make or handle charcoal once the law became active. Producers/importers of over 1,000 tonnes of charcoal were asked to pre-register their intentions by 2010. For those in business making 10–100 tonnes, pre-registration was required by 2013; for burners making under 10 tonnes a year the deadline is 2018.

In summary, there is no exemption from REACH legislation for charcoal makers planning to make between 1 and 100 tonnes of charcoal or biochar a year. If you plan to make a large quantity of charcoal, you must seriously consider this, as your customers may have concerns about the charcoal-maker's compliance with the law.

In 2015, in correspondence with Darren Hopkins of the British Biochar Foundation, who has gone to great lengths to get clarity on this subject, the H&SE HM Inspector of Health and Safety, Mike Potts, stated that in relation to biochar production there was:

> ...negligible risk to human health and the environment associated with this substance (bio-

---

Great Britain's Registered Charcoal Producers

The British charcoal industry (manufacture, imports and sales) has four big players: Rectella, Big K, CPL and Direct Charcoal. All four companies import charcoal (Rectella from Namibia, Big K from South Africa), which is often repackaged in the UK into other products such as barbecue foil trays, or various medicinal products. 110 companies, including the biggest four companies from Britain, have submitted a joint registration, and in addition to that thirty-two companies are registered EU wide at 1,000–10,000 tonnes, sixty-eight at 100–1,000 tonnes, five at 10–100 tonnes and just one at 1–10; many of these companies are Greek or Spanish. As we have seen in previous chapters there are other manufacturers producing considerable quantities of charcoal in the UK, some of whom may not yet be registered.

> char), I consider this matter to be a low priority for investigation. As our current case load is such that our resources are fully deployed on high priority, risk-based casework, I do not intend to undertake any investigation against you in regard to this matter, as such a response would not be consistent with our obligations as a regulator.

REGULATIONS REGARDING PACKAGING AND MARKETING CHARCOAL

The British Standard BS EN 1860–2:2005 covers 'Appliances, solid fuels and firelighters for barbecuing. Barbecue charcoal and barbecue charcoal briquettes. Requirements and test methods'.

This states that granulation size should be as follows:

0–150mm:
No more than 10 per cent may exceed 80mm in size

At least 80 per cent shall be greater than 20mm
0–10mm shall not exceed 7 per cent

Riddling the charcoal over a 10mm mesh removes most of the small particles, and those larger than 80mm can either be broken up or put aside for restaurant-grade charcoal. Care should always be taken when handling and stacking charcoal bags as it can break down, ending up as dust in the bottom of the bags, which may be great for the garden but is not much good to burn.

Weighing Charcoal

Charcoal and biochar sold by weight or volume is governed by the 'The Weights and Measures (Packaged Goods) Regulations 2006'. This states in clause 3.1:

> *3.—(1) Subject to paragraphs (2) to (6), these Regulations apply to:*
> *(a) packages intended for sale in constant unit nominal quantities which are:*
> *(i) equal to values predetermined by the packer;*
> *(ii) expressed in units of weight or volume; and*
> *(iii) of not less than 5 grams or 5 millilitres and not more than 25 kilograms or 25 litres;*
> *(b) outer containers.*

This implies that if you don't predetermine the value and or express it in units of weight or volume then the regulations do not apply. If you do want to state the weight or volume on the bag then you have a 'Duty to comply with the "three packers" rules':

> *4. – (1) It shall be the duty of the packer or importer of packages to ensure that they are made up in such a way as to satisfy the following rules —*
> *(a) the contents of the packages shall be not less on average than the nominal quantity;*
> *(b) the proportion of packages having a negative error greater than the tolerable negative error shall be sufficiently small for batches of packages to satisfy the requirements specified in Schedule 2;*
> *(c) no package shall have a negative error greater than twice the tolerable negative error.*

Platform weighing scales.

> *(2) Compliance with the rules in paragraphs (1) (a) and (b) shall be determined by the reference test.*

The reference test is worded thus:

> *1. REQUIREMENTS FOR MEASURING THE ACTUAL CONTENTS OF PACKAGES*
> *1.1 The actual contents of packages may be measured directly by means of weighing instruments or volumetric instruments or, in the case of liquids, indirectly, by weighing the packed product and measuring its density.*
> *1.2 In all operations for checking quantities of products expressed in units of volume, the value employed for the actual contents shall be measured at, or corrected to, a temperature of 20°C,*

whatever the temperature at which packaging or checking is carried out. However, this rule shall not apply to deep frozen or frozen products, the quantity of which is expressed in units of volume.

1.3 Irrespective of the method used, the error made in measuring the actual contents of a package shall not exceed one-fifth of the tolerable negative error for the nominal quantity in the package.

Tolerable error

500g to 1,000g + or –   15g or ml

1kg to 10kg   + or –   1.5%

The legislation states that equipment used should be 'suitable for the use to which it is put'. Whilst this is not very exact, in the event of a charcoal maker being challenged and found to have been weighing bags on inaccurate (and unsuitable) bathroom scales, there may be a legal case to answer. A spring balance is a better option, or a balance with weights, which can be checked for accuracy. Remember to check that the mechanism in the weighing scales used is not getting clogged up with charcoal dust. For volume measurement a 0.7gal (3ltr) bucket is good, while quality control should ensure that everyone employed in filling the bags, fills to the standard level.

CERTIFICATION SCHEMES TO VALIDATE SUSTAINABILITY CLAIMS

Certification of Woodlands

There are two main certification schemes for woodlands, forests and wood products, one run by the Forest Stewardship Council (FSC) (international but UK-based), the other by Programme for the Endorsement of Forest Certification (PEFC) (international but European based).

Forest Stewardship Council (FSC)

The FSC describes itself as 'an international, non-governmental organization dedicated to promoting responsible management of the world's forests'.

FSC Ten Principles for Responsible Forest Management

1. Compliance with laws and FSC Principles – to comply with all laws, regulations, treaties, conventions and agreements, together with all FSC principles and criteria.
2. Tenure and use rights and responsibilities – to define, document and legally establish long-term tenure and use rights.
3. Indigenous peoples' rights – to identify and uphold indigenous peoples' rights of ownership and use of land and resources.
4. Community relations and workers' rights – to maintain or enhance forest workers' and local communities' social and economic well-being.
5. Benefits from the forest – to maintain or enhance long-term economic, social and environmental benefits from the forest.
6. Environmental impact – to maintain or restore the ecosystem, its biodiversity, resources and landscapes.
7. Management plan – to have a management plan implemented, monitored and documented.
8. Monitoring and assessment – to demonstrate progress towards management objectives.
9. Maintenance of high conservation value forests – to maintain or enhance the attributes that define such forests.
10. Plantations – to plan and manage plantations in accordance with FSC principles and criteria.

The scheme is based on the 'Ten principles for good forest management' (*see* box), which is assessed by independently accredited bodies who visit and certify that woodlands are being managed in an 'environmentally appropriate, socially beneficial and economically viable manner'. There is also a 'chain of custody' set up, whereby each stage in the production of a wood product – from harvesting, moving the product, milling or processing and then manufacturing (in the case of charcoal) – has to meet the required standards and be registered with the FSC.

PEFC: Programme for the Endorsement of Forest Certification

The PEFC is described as 'the world's leading forest certification organization', and 'an international, non-profit, non-governmental organization dedicated to promoting sustainable forest management, the PEFC is the certification system of choice for small forest owners'. PEFC also runs a chain of custody system and has independent monitoring.

There is not much between the two schemes, but the main difference is in the 'not for profit' status of PEFC. Both FSC and PEFC have collaborated and participated in the United Kingdom Woodland Assurance Scheme (UKWAS) steering group.

United Kingdom Woodland Assurance Scheme (UKWAS)

UKWAS is where the two main certification schemes have come together to form the 'gold standard' of woodland management in Britain. The UKWAS steering group comprises a range of stakeholders in the UK forest industry, covering the three main elements of good forest management, taking account of the economic, social and environmental role of woodlands. It has published extensive guidelines, which are regularly updated with the latest best practice for sustainable forest management.

Grown in Britain

'Grown in Britain' (GiB) is an initiative set up in 2010, to attempt to boost the UK's home-grown forest industries. Over 60 per cent of timber used in this country is imported, including almost 90 per cent of charcoal. Grown in

'Grown in Britain' logo.

Britain supports locally sourced timber from sustainable woodland management. They use their clear logo to maximum advantage by licensing those who are part of the home-grown timber supply chain to use their logo on packaging and products. The assurance scheme provides independent verification that the timber is 'locally produced' timber from well managed woodlands. Buying Grown in Britain products gives our forests and woodlands a sustainable future, promotes employment and local economies, and cuts 'wood miles'. A group scheme has been developed for coppice workers supported by the National Coppice Federation (NCFed) and the Kent Coppice Group, to enable those who work on a smaller scale to badge their produce.

The Fairtrade Initiative

The Fairtrade initiative also has a certification system, and it would be good to see Fairtrade principles extended more widely to include imported charcoal. There is a development by Traidcraft, the company that deals in Fairtrade goods, in collaboration with Rectella (as noted above, one of the bigger charcoal importers in Britain) and Co-operative supermarkets, to import Fairtrade charcoal from Namibia. The charcoal is reportedly made from acacia trees, which, being thorny and unpalatable to browsing animals, become dominant on farmland. Removal improves agricultural productivity. The timber source is FSC certified. The company states: 'The Fair Trade charcoal will benefit around 1,000 producers and will mean they receive a guaranteed contract, payment of around three times the minimum wage, and free equipment such as charcoal kilns.'

BIOCHAR CERTIFICATION

Three biochar standards are relevant in Britain, which lay down quite strict rules regarding environmental standards and quality issues. (For more in-depth information on biochar, see Chapter 9.)

British Biochar Foundation

The British Biochar Foundation has published a 'Biochar Quality Mandate', a detailed document that covers all aspects of biochar making, marketing and use. Its definition of biochar is as follows:

> Biochar is a solid material obtained from the thermochemical conversion of sustainably sourced biomass in an oxygen-limited environment, using clean production processes and which is used for any purpose that does not involve its rapid mineralization to CO2.

The quality mandate divides biochar into two different bands: 'standard grade' and 'high grade'. The main difference between these two grades is in the level of permitted limits of toxic substances within the biochar. In simple terms, the biochar should be safe to apply to soil, and have a minimum of 10 per cent stable organic carbon, and also a minimum of 30 per cent of the carbon content of the original feedstock. The net carbon balance should be positive, including all elements of the production line.

European Biochar Certificate (EBC)

This certification was developed and first published in 2012 to provide a European industry standard for biochar. A certificate has now been developed for small-scale producers producing less than 20 tonnes a year. The EBC defines biochar as follows:

> A heterogeneous substance rich in aromatic carbon and minerals. It is produced by pyrolysis of sustainably obtained biomass under controlled conditions with clean technology, and is used for any purpose that does not involve its rapid mineralization to CO2 and may eventually become a soil amendment.

The EBC certificate ensures that producers who meet the standard comply with the following:

• Sustainable provision and production of biomass feedstock

A lovely vegetable garden created with biochar.

- Energy efficient, low emission pyrolysis technique
- Biochar quality – low contaminant level
- Low hazard use and application of biochar (EBC 2012)

International Biochar Initiative

Although developed independently, the standards set by the International Biochar Initiative (IBI) (2012) are remarkably similar, with a simple definition of biochar as 'A solid material obtained from thermochemical conversion of biomass in an oxygen-limited environment' (IBI, 2012).

Of the three organizations, the IBI has the loosest definition of biochar, which could in fact be applied to any form of charcoal. In contrast, both EBC and BBF specify 'clean technology' as being a key characteristic, which is an attempt to exclude more polluting manufacturing methods of earth burns and kilns. Both organizations specify that biochar as a substance should not involve 'rapid mineralization to CO_2 or being burnt, as is the destiny of barbecue charcoal'. Only the EBC mentions that the substance 'may be used as a soil amendment'.

Biochar Risk Assessment Framework

Set up at the UK Biochar Research Centre in Edinburgh, but with collaboration from a number of interested partners including the National Farmers Union and the Environment Agency, the Biochar Risk Assessment Framework states:

> *The purpose of this project is to identify and analyse the potential risks of biochar production from different types of biomass feedstocks (bio-energy crops, agroforestry residues, biodegradable waste, etc.) and its deployment in farming*

This group intends to develop methodologies for using biochar that will comply with UK and EU regulations.

Chapter 8
Cooking With Charcoal

An estimated 10 million tonnes of charcoal are produced annually in the world, and an estimated 50,000 tonnes of this is used every year in Britain for barbecue fuel. What makes charcoal so popular as a fuel? It has few inorganic impurities, is almost smokeless, and has a high ratio of carbon to ash, so it gives a fairly even, consistent burn temperature. The key quality of charcoal, as compared to coal or coal-derived products, is the low sulphur content, so the odours emitted are almost non-existent or at best are aromatic.

CHARCOAL AS A COOKING FUEL

Charcoal is used, by choice or necessity, for domestic cooking across the globe, especially where the cost of fossil fuels is prohibitive and electricity supplies are unreliable. As economies grow and populations have more disposable income, there seems to be a desire to remain connected to these more basic survival skills, which is demonstrated in our enthusiasm for 'al fresco' cooking. This may be the

A charcoal cooker for mass catering.

reason for the rise of the domestic barbecue, and a range of charcoal and adulterated charcoal products to fuel them. This has created a massive import market in the wealthier nations, where thousands of tonnes are bought annually for recreational cooking from less developed countries where the fuel is cheap.

2015 trade figures for charcoal show that Britain imported 45,139 tonnes of charcoal from non-EU countries at a value of £18 million. Sub-Saharan Africa represents the largest source of imported charcoal with nearly 25,789 tonnes at a cost of £11 million, or £426/tonne. The second largest source of imported charcoal is Latin America's 12,500 tonnes at a value of £4.7 million or £376/tonne. Further charcoal comes in lesser quantities from Asia and the Middle East.

However, the demand for charcoal in the UK has dropped over the past twenty years: in 1996 imports were 55,000 tonnes at an average price of £182/tonne. Perhaps this shift is due to the rise in the gas barbecue, or perhaps there is less patience with the vagaries of the British climate, which makes a culture of cooking 'al fresco' a chancy business. There is, though, a growing market for British charcoal produced by UK burners making 1–300 tonnes of charcoal a year. UK production is estimated at about 8–10 per cent of the total sold, or 7,000 tonnes, with an average wholesale value of £1,000/tonne.

Charcoal, whether imported into Britain or manufactured here, either from native or imported timber feedstock, is predominately used for barbecues. It may be sold either as lumpwood charcoal or as briquettes, which are a composite product made from charcoal dust and other substances.

Lumpwood Charcoal

Lumpwood charcoal is converted from whole timber or branchwood and has no additives, just the carbon and any residual volatiles that were not driven out, and some ash (in fact there is very little of either of these in well made, high carbon-content charcoal). Equally, it can contain varying percentages of volatiles depending on the temperature to which the timber was

---

Charcoal Production in Kenya and Malawi

Charcoal Production in Kenya

> *Charcoal is a key bioenergy resource and source of energy in Kenya, providing 82 per cent of urban and 34 per cent of rural household energy, and employment and income for over 700,000 people who support over 2 million dependants.*

Demand for charcoal is fast increasing due to population growth, increased urbanization, and the development of cottage industries. Between 2000 and 2009, the Government of Kenya formulated policies and legislation on charcoal production (tree growing and wood conversion to charcoal), transportation, trade and utilization.

(Kenya Charcoal Regulations Pocketbook)

Charcoal Production in Malawi

> *Charcoal production is regarded as one of the major contributing factors to loss of biodiversity and clearing of forests in the region. In the Shire River Basin a large number of people depend for their livelihood on the burning and selling of charcoal. Fuel consumption of charcoal in nearby Blantyre City alone is estimated to be around 1.06 million tonnes per year, and this is equivalent to cutting 36ha of trees each day of the year. Charcoal production and the opening up of land for agriculture are the major contributing factors to deforestation, hence environmental degradation needs to be addressed urgently if the goals of the MGDS (Millennium Development Goals) are to be realized.*

Unfortunately, as charcoal production remains quite traditional, it tends to result in inefficient wood conversion to charcoal, thereby exacerbating the problem.

From: 'Review of Policies and Regulations on Charcoal and how to Promote a Systems Approach to Sustainable Charcoal Production and use In Malawi': United Nations Development Programme.

A comparative table for lumpwood charcoal and briquettes

	Pros	Cons
Briquettes	Uniform product; even burn. Recycles a waste product (sawdust). Cooler burn. Can contain additives that make it easier to light.	Contains mineral carbon other than wood that may impart unwanted flavours. Has a high ash content which can clog up equipment when using. Made in centralised manufacturing units requiring the transportation of materials and increased distances for delivery.
Lumpwood	100 per cent wood product with potential for individual flavour characteristics depending on species used. Can be used in small-scale production units close to timber source and to markets. Hotter burn and gets to temperature more quickly. High quality lumpwood charcoal lights very easily.	Can be of inconsistent quality. Breaks down in the bags if roughly handled.

taken in the manufacturing process. A high volatile content can impart more flavour as the volatiles will be emitted as smoke, but the balance between heat and flavour is quite delicate. Dense charcoal equals heavy charcoal and can be poor value for money when purchased by weight.

Briquettes

Briquettes were developed in the 1920s in America by Henry Ford, as a way of using the waste products of his car-building business (when bodywork was mainly wooden). Nowadays briquettes utilize by-products of the timber and paper industries, including bark, twigs and sawdust. The waste is chopped to a uniform grade and charred in a retort before being mixed with a 'binder' and/or a 'filler' and formed into briquettes. Occasionally they are made of charcoal dust that is too fine to burn on its own. The resulting fuel is denser and heavier than high carbon-content lumpwood charcoal, and its main positive characteristic is that it is uniform. Whether briquettes burn as well as, or better or worse than lumpwood charcoal may depend on the manufacturer and the quality of the raw material and the additives.

Lumpwood Charcoal Versus Briquettes

Briquette Ingredients

The basic ingredient of briquettes is wood charcoal from timber and paper manufacturing waste (sawdust). 'Binders' are substances that act as an adhesive for the particles; these include starch from corn, wheat, rice or potatoes. 'Fillers', or additives to briquettes, include the following:

Limestone: Gives the briquettes a white coating when they are burning and so acts as a visual clue to the fuel being ready to cook on.

Anthracite: A mineral coal dust which increases the heat of the burn.

Mineral char: A form of 'young' coal which is mined but is brown in colour and soft in texture, a kind of halfway-house between peat and coal; it may have an odour of decomposition.

Other additives may be:

Borax: A mineral, but acts to allow the briquettes to be released from their moulds without sticking.

Sodium nitrate: Can be used as an aid to ignition.

Dried but uncharred sawdust: Also aids combustion as it has a low water content but is high in volatile oils contained in the wood (and may produce smoke on ignition).

Restaurant-Grade Charcoal

Restaurant-grade charcoal is the same as barbecue charcoal but is screened to favour larger pieces, ideally 40–150mm, which burn slowly and last longer per charge. Retort-made charcoal is absolutely ideal for this use, as the wood tends to remain more structurally intact and in larger pieces even when heated to 450°C.

Some types of restaurant-grade charcoal have a lower carbon content and a higher volatiles content. Chefs appear to love its smoky nature and the associated flavours the smoke gives the food.

'Par char', or charcoal 'brown ends', can be used for cooking too, as long as there has been a partial conversion. Brown ends are often discarded as a waste product by the charcoal burners – they are usually used to start the next burn – as they do not conform to the desired carbon content for quality barbecue fuel.

Torrefied Wood

Wood is described as torrefied when the wood waste material is taken up to 300°C, so that many of the volatiles remain and the percentage of carbon is lower than that of charcoal. It has the benefit of having a low

Picture of briquettes from France.

Restaurant-grade charcoal. JIM BETTLE

Simple cooking.

Mark Parr, Charcoal Producer for the Restaurant Trade

As a teenager, Mark Parr would go and watch the Arundel charcoal burners in Kent, who used old metal ring kilns; he was fascinated by the process of charcoal making. After leaving school he went to art college, later pursuing a career in design, working for top designers including Cath Kidston. While living in Crystal Palace, London, he realized how tricky it was to source firewood, and decided to set up a firewood business, starting out with £180 and a pallet of logs which he sold to his neighbours. He soon developed a website and a regular blog about wood and woody things, which really opened up his markets and led to some 'A list' customers.

At this time he made a fortuitous connection with Lord Falmouth's Mereworth estate in Kent, which provides the company with a huge timber resource of mainly sweet chestnut and mixed coppice. He was encouraged to take up charcoal production for the restaurant trade – he already had connections within this sphere, and in another life might have been a chef himself. Since 2012 he has built up the London Log Company, which now supplies 300 restaurants in London and has a client-base of 10,000 on the books, turning over £1.4 million in the last financial year.

moisture content and high volatile content, so is good to burn and lighter to transport than wood, but is not as 'smokeless' as charcoal, and has a cooler burn.

Gourmet Charcoal

Often charcoal is sold as 'single species' charcoal, each different tree species having a slightly (or marked) different chemical make-up, conferring a different flavour or quality to the product. Mark Parr suggests running a trial with different types of charcoal on some plain chicken breast, prawns or a simple fish dish to detect the variation in flavour. The box [*see*

opposite page] describes how a wonderful range of flavours can be achieved in the food, from the tangy metallic notes achieved when cooking with oak charcoal, to the softer, buttery flavours with hazel charcoal. Chefs are trained to detect the range of aromatic qualities from dark to light, sharp to sweet. This attention to detail has given the British charcoal industry a boost in recent years. The key element, noted in Chapter 5, has been the introduction of twin-chambered retorts such as the Pressvess and the Oxford retort, in which it is possible to burn relatively green fresh wood. The water content is reduced in the drying phase of the burn, maintaining the 'fruit notes' that have not been oxidized as they might in air-dried timber.

THE PLEASURE OF BARBECUING

Types of Barbecue Appliance

Bricks or Stones

A wide range of barbecue appliances is available on the market – and an efficient barbecue can

Qualities of Charcoal from Different Species

Oak
English production charcoal:

- Deep and rich 'beurre noisette' notes
- High mineral spike to finish, underscored with vanilla, tobacco and peat. Clear flame and smoke character
- Highly carbonized with distinctive aromatic notes

Beech

- Clear and light sweetness, with a clean flame and smoke character
- Contains a natural polysaccharide within the wood (xylitol), which comes through in the burn notes
- Highly carbonized with distinctive aromatic notes

Silver Birch
English woodland charcoal:

- Zesty, citric, with rich base mineral notes.
- Clean flame and smoke character. Usually found only in Finland
- Highly carbonized with distinctive heady-sweet-oily aromatic notes

Wild Cherry
English wild woodland charcoal: we only get access to the odd thread of this wood, but it is here
- Deep, strong, rich and 'of the earth'

- Not overpowering, but a character all of its own: think Middle Earth and Dark Magic. Clean flame and character

Sweet Chestnut
English woodland charcoal:

- Clean and distinctive, almost fruit-rum note and an ultra-clean flame.
Highly carbonized with distinctive, clean, aromatic notes.

Hazel
English woodland charcoal:

- Burns to a unique, pure white colour
- Light smoke, with crème caramel/nutty aromatic characteristics
- Highly carbonized with good, clean-tasting notes

Apple
English woodland charcoal:

- This is probably the hardest, maddest, loveliest wood to carbonize. It takes the longest to prepare and fire, as the wood is as hard as nails and weighs a ton. But the flavour is worth the effort
- Highly carbonized with rich, smoky fruit and distinctive aromatic notes.

From: The London Log Company website http://thelondonlog-company.blogspot.co.uk

be improvised at home, the simplest of which is a couple of bricks or rocks laid on the ground to take a grill. If you want to leave no trace of where you have been, use a spade to peel back the turf before building it, and afterwards when the ground is cool, this can be firmed back into place, with no one any the wiser that you have been cooking there.

A car wheel-rim also provides a makeshift barbecue. Fill the hole in the base with some larger chunks of charcoal, or place some wire netting across it to prevent the fuel from slipping through the bottom. To provide a draught, lift the wheel-rim a little on some flat stones or pieces of wood. Old oven shelves can be used to grill the food on.

All these options only work if you don't mind getting down to ground level to cook! However, raising the firebox up to a handy height is preferable, to minimize back trouble. For those keen on barbecued food, the best option is to buy (or make) an appliance on legs.

The image below shows a classic half oil-drum barbecue: this one has a nice touch in that you can invert the drum to prevent it filling with rainwater. The barrel barbecue is improved with a fine-mesh shelf, inserted at about 1in (20mm) depth to hold the charcoal, otherwise the fire is too far down to be effective.

A barbecue with a hinged lid will keep the firebox dry and ready to light at short notice. A lid can be used to transform the barbecue into a smoker, or to keep food moist. Another useful refinement is to be able to raise or lower the grill, raising it to avoid burning, or lowering it to use up the last dying heat of the charcoal. Having the appliance on wheels will make it easier to move it around.

Lighting a Barbecue

The simplest way to light charcoal is with paper and/or firelighters broken up and spread amongst the charcoal. Wax-based lighters are usually paraffin wax, so to avoid contamination they need to have burned away completely, before starting to cook. Try to find eco-firelighters, which are resin-based and don't impart an aroma of paraffin to the barbecue. If lighter fuel is needed, use a different type of charcoal, because higher-quality charcoal is drier, burns more readily, and does not need extra combustibles to ignite properly.

Some people would not attempt to light a barbecue without the aid of a 'chimney charcoal starter lighter kit', as shown on page 136. This keeps the charcoal together and creates a draw as the heat rises with an air flow from underneath. The fire is soon set with dry newspaper in the bottom, a good layer of charcoal on top, and a match or flint to ignite it. Once it is going

A half-barrel barbecue on legs.

General Safety Information Given by the Fire Service

- Make sure your barbecue is in good working order
- Ensure the barbecue is on a flat site, well away from a shed, trees or shrubs
- Keep children, garden games and pets well away from the cooking area
- Never leave the barbecue unattended
- Keep a bucket of water or sand nearby for emergencies
- Ensure the barbecue is cool before attempting to move it

Charcoal barbecues:
- Use only enough charcoal to cover the base to a depth of about 2in (50mm)
- Only use recognized firelighters or starter fuel, and only on cold coals – use the minimum necessary, and never use petrol
- Never put hot ashes straight into a dustbin or wheelie-bin as they could melt the plastic and cause a fire

From: http://www.fireservice.co.uk

Barbecue with a lid.

well, the hot coals can be spread about the fuel box of your barbecue appliance.

The chimney lighter can be supplied with a handle for ease of use. Alternatively, a chimney lighter can be improvised from a large metal food can pierced with some holes. Remember to wear a pair of fireproof gloves to handle it safely.

Barbecue Lighter Fuel

Many types of lighter fuel claim to be odourless, but most are made from paraffin or kerosene (mineral oil distilled from coal, and very toxic). Barbecues lit with this kind of fuel typically fill the garden, and neighbouring gardens, with rather heavy, paraffin-scented smoke, which takes some time to abate. A more pleasant alternative is ethanol (alcohol) gel, marketed as 'eco' and non-toxic – but as noted above, the best option is to buy better quality charcoal, dispensing altogether with the need for these kinds of lighter fuel.

ABOVE LEFT:
A chimney lighter.

LEFT: *Eco firelighters.*

Cooking on Charcoal

When the charcoal is lit and going well you can soon spread the lumps, adding more to make an even layer over the base grill of the appliance. Received wisdom is to wait before starting to cook until there is a white ash coating on the coals – however, this is less apparent in good quality lumpwood charcoal, as it has a low ash content. Briquettes are more likely to form white ash, as limestone is often added to create this white ash effect. Usually it takes about ten to fifteen minutes for the temperature to reach cooking heat, and you can make a start with the items that will take longest to cook.

The portable barbecue (see below) has a lid for containing steam and smoke, which helps to increase the cooking options and quality; the wheels are useful for moving the barbecue

A portable barbecue
on wheels.

around, though it tends to be top heavy. Here it is being used to cater for a family party of about twelve people. One charge of charcoal is usually enough, but refuelling will be necessary if the event lasts more than an hour. Each time more charcoal is added, wait ten to fifteen minutes for the new fuel to ignite and get up to temperature.

Barbecue Trouble-Shooting

Food cooked on a too-hot barbecue will be burnt, cooking too fast on the outside but still raw on the inside. Solutions include:

• Use less fuel on the barbecue, it never needs much, just a layer on the bottom of the fire pan
• Raise the grill up away from the fire; a height adjustable grill is helpful to find the right distance from the coals for perfect cooking
• Avoid adding too much extra fat to the food, which can produce too much smoking, particularly if cooking meat, as the natural fat in meat melts and drops on to the coals, creating flame and smoke
• This can be partly avoided by using a lid to reduce the oxygen, or wrap the food in foil to steam rather than bake, retaining the natural juices and fats

Barbecue Recipes

Home-made burgers can be made very simply using fresh beef chuck steak put through a coarse mincer, and seasoned with salt and pepper. The mix should stick together and form burger shapes for cooking on the barbecue. Cook for four minutes on either side for a rare burger, or six minutes each side if required. Sausages are always popular and easy to cook, and are available in a huge range of flavours and sizes.

Be sure to cook all meat products right through, especially pork or chicken-based products, which must be thoroughly cooked. Kebabs or skewers of meat and vegetables are an easy way to turn and cook food evenly, but wooden skewers should be soaked in water beforehand to ensure they don't burn.

Cooking for Vegetarians

Barbecues are not always natural territory for vegetarians – typically barbecues tend to be heavily meat focused – but delicious vegetarian or vegan food can be cooked on a separate grill. Vegetable kebabs can include mixed vegetables such as mushroom, peppers, tomatoes and aubergine, and chunks of halloumi cheese. Some care is required to prevent the cheese from sticking to the grill and falling off the skewer, but is very tasty if cooked well with a light basting of olive oil to prevent the vegetables from drying out. Sweetcorn and asparagus spears can also be roasted directly on the grill.

Home Appliances for Smoking Meat and Fish

Home appliances are available to purchase, which use charcoal fuel for smoking meats and fish; alternatively a DIY smoker can be con-

An Excellent Smoked Food Experience

One of the very best smoked food experiences the author recalls was some years ago, courtesy of Iain Spinks of Arbroath Smokies at the Royal Highland Show. Iain arrived, leapt out of his van, and started to excavate a pit on the showground big enough to insert a half oak barrel. The inside of the barrel was lined with slate to protect the wood from burning, and a fire was started in the barrel using charcoal. Once properly alight, it was stoked with oak and beech logs, and racks of haddock fillets tied in pairs, and as many as would fit strung over a rail which sat on the rim of the barrel. The whole contraption was covered in wet hessian sacking and left to cook and smoke for thirty to forty minutes, and the result: the most delicious hot smoked fish.

Recipe for Caroline Bingley's Yummy Kebabs

1 tbsp olive oil
1 tbsp honey
2 garlic cloves roughly chopped
2 tbsp soya sauce
2 chicken breasts, or cubed packet of halloumi
1 red pepper
1 courgette
1 onion
Handful of cherry tomatoes
Squeeze of lemon juice

Cut the chicken and veg into 1in (2.5cm) chunks, and place in a large bowl or pan. Mix marinade ingredients together with the chicken and veg and marinade for at least four hours, preferably overnight.

Using skewers, arrange the chicken and vegetables alternately on the skewers, and cook on the barbecue, turning frequently until all the chicken pieces are cooked through with no raw pink bits left – around 10 minutes on a medium barbecue. Enjoy with corn on the cob and a range of salads. The chicken can be substituted with halloumi or aubergine if preferred.

Chicken kebabs. CAROLINE BINGLEY

Mass catering in France for the European Charcoal-burners Convention.

Lembacherflackstaaner-kohlebrannergrumbare-roschtwurst.

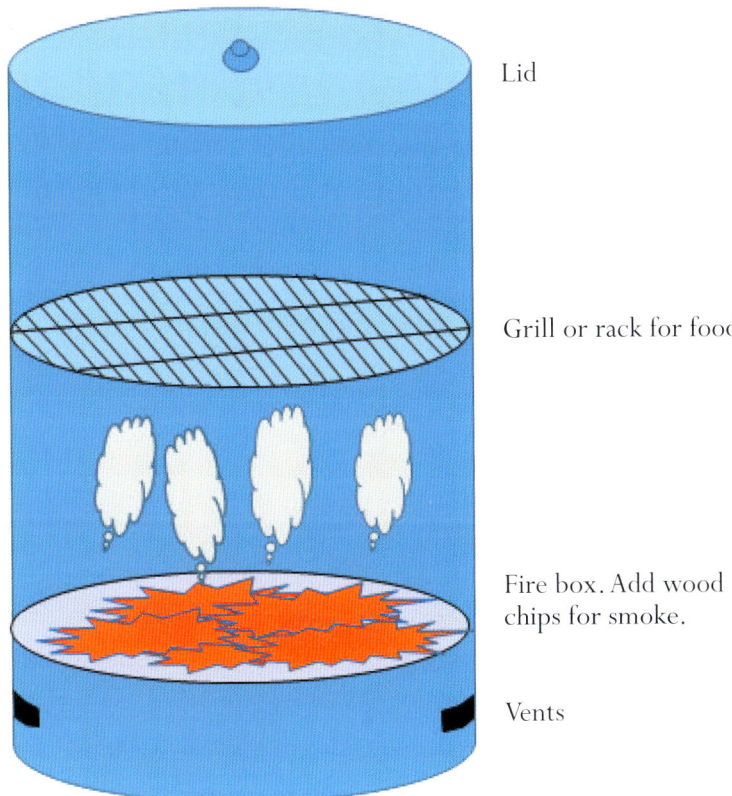

Lid

Grill or rack for food

Fire box. Add wood chips for smoke.

Vents

A basic smoker.

structed. The main requirement is a firebox and a smoking chamber, with racks on which to place or hang the food to be smoked.

A smoker is an enclosed firebox chamber ventilated from below to ensure sufficient oxygen for the fire to burn. Above the chamber are racks for the food and a lid to keep heat and smoke in, with vents to create a through-flow of smoke.

Mass Catering

So long as you are well organized, there is no excuse to resort to gas grills when catering for large numbers. At the European Charcoalburners Convention up to five hundred people were catered for: ham hock wrapped in foil, and the delicious legendary sausage Lembacherflack-staanerkohlebrannergrumbareroschtwurst – all cooked on the barbecue.

Chapter 9
Biochar and Carbon Sequestration

The global phenomenon of biochar production for use as a soil amendment has passionate exponents, and a few equally passionate detractors. This chapter attempts to unpick some of the arguments, and to examine the debate about the use of small particles of charcoal ('fines') – or 'biochar' – as a soil improver, which is promoted both to enhance fertility in soil, and potentially play a role in carbon capture or 'sequestration'. Biochar is a recent term for an ancient practice where small particles of charcoal are added to improve or 'amend' soil. Inspired by studies of the Terra Preta soils of South America, awareness has grown of the potential of biochar for improved yields in agriculture and horticulture.

The International Biochar Initiative (IBI) was formed in 2009 to coordinate the research into the potential for biochar, with a focus on four key areas of biochar usage:

- Incorporation of charcoal into agricultural systems to improve plant growth
- The locking of carbon into the ground (carbon sequestration) to reduce carbon dioxide in the atmosphere
- Conversion of waste materials into carbon
- The utilization of charcoal as a by-product of biofuels

Of these, the fourth point has raised alarm bells for some, as there is an anxiety that uncontrolled market forces will encourage over-production of biomass for biofuels and the charcoal produced as a by-product of biofuel. This may have a detrimental effect on natural ecosystems and food production, as has already been seen and documented by organizations such as 'Biofuelwatch'. The concern has led to some close scrutiny of the trial results produced on bio-

Pumpkins grown in biochar. JASON SMITH

char; some have been found to have a neutral result, some a negative result. Many trials, however, have proved positive, and we will attempt to unravel the possible reasons for these mixed results.

BIOCHAR IS BORN

Terra Preta, or Dark Earths

Throughout the twentieth century interest grew in the 'Terra Preta' or 'dark earths' of the Amazon Basin. It was observed that these exceptionally fertile soils contained, amongst other things, a high carbon content. The late Dutch soil scientist Wim Sombroek (1934–2003), Secretary General of the International Society of Soil Science from 1978 to 1990, was a leading expert on this subject, whose life work was studying the Terra Preta, or 'dark soils', found in the Amazon basin and elsewhere. He suggested that the use of charcoal as a soil improver (with added nutrients) enabled an entire civilization to prosper in South America between 450 and 950 years ago.

Studies of how these soils were formed suggest that our current anxiety about deforestation of the Amazon through 'slash and burn' agriculture may be mitigated by the likelihood that 'slash and burn' agriculture was preceded by a system which has been characterized as a highly productive 'slash and char' system, which has relevance for us today. Sombroek was ahead of his time in recognizing the potential for combating climate change through the sequestering of carbon in the form of charcoal in soils.

The Terra Preta Nova Group

The twenty-first century has seen an acceleration of interest in adding charcoal to soils. Research conducted by the Terra Preta Nova Group, who published *Amazonian Dark Earths: Wim Sombroek's Vision*, edited by W. I. Woods in 2009, includes work by, amongst others, W. I. Teixeria, Christoph Steiner and Johannes Lehmann of Cornell University New York.

The Ithaka Institute

The Ithaka Institute for Carbon Intelligence, set up in 2008 by Hans Peter Schmidt in Switzerland, has offices in Germany, Australia, Nepal and the United States. The Ithaka Institute is a leading organization promoting research into biochar, and in particular championing open access technology to make biochar production cleaner and more affordable, through the development of their Kon-tiki kiln. They produce the informative *Biochar Journal*, edited by Kathleen Draper who is also Director of the American branch.

In Britain, the UK Biochar Research Centre at the University of Edinburgh linked up with three Canadian universities to look at the potential of biochar for carbon sequestration. This is discussed later in the chapter.

Organizations Promoting Biochar

The concept of creating carbon and storing it back in the earth soon gained many enthusiastic proponents, and led to the forming of the International Biochar Initiative (IBI), which reports that 380 papers on biochar worldwide were published in 2013 alone. The European Biochar Foundation (EBF), formed in 2012, and the British Biochar Foundation (BBF), which held two national meetings in 2013 and 2014, have both raised the profile and set standards for biochar. As seen in Chapter 6, they all have a fairly similar approach to the subject, and are concerned to distance themselves from inefficient, polluting methods of charcoal production and from the potential ecological down side of mass crop production for feedstock, to the detriment of other crops and the natural environment.

The United States Biochar Initiative (USBI) describes biochar as:

> *A zero-waste solution — biochar is a fine-grained charcoal made by pyrolysis, the process of heating bio-mass (wood, manure, crop residues, solid waste, etc.) with limited to no oxygen in a specially designed furnace capturing all emissions, gases and oils for reuse as energy.*

This approach is somewhat divergent from the European approach, which has distanced itself from the link with biofuels.

Charcoal Fines

Earth burns, metal kilns or retorts will all produce 'fines' as a by-product of making lumpwood barbecue charcoal. The smaller particles screened out through 10–12mm mesh are known as charcoal fines. Often discarded in heaps near to the kilns, this undervalued by-product of charcoal production has quietly been gathering devotees, and the innovative and entrepreneurial burner has both winkled out new markets and revived old markets. Now, the buoyant demand for fines is being met by specialist burners and crushing machines, and, of course, resorts to cheaper imports. With the adoption of the term 'biochar', charcoal fines have found a renewed purpose in life.

A Fines Case Study

When the author was annually producing half a tonne of barbecue charcoal a week between March and August, there was not much call for horticultural charcoal. A neighbouring burner won a contract with Edinburgh's Royal Botanic Gardens, which already knew the merits of charcoal, particularly for orchid growing, and would buy it by the tonne annually. Another colleague had special small bags printed, and would sell it by the kilo at shows and occasionally to local garden shops. The value of this was somewhat theoretical, and perhaps rather too much trouble when you had your head down producing barbecue charcoal.

Between starting burning in 1994 and finally finishing in 2010, all my 'fines' were saved, bagged and stored as what became jokingly known as 'the pension pot'. An exciting moment occurred in 2009 when colleague Ian

Fines stored in a dumpy sack.

The Benefits of Biochar in Forestry and Ground Reclamation

David Hutchinson of the Yorkshire Charcoal Company has a background in horticulture, but began charcoal burning in 1987 from his site near Scarborough in North Yorkshire. He pioneered early trials of the use of charcoal fines by the Forestry Commission (FC). In 1992 a two-year trial with nursery-grown conifers was almost unbelievably successful, so the trial was repeated – again with a very positive result. A year later, a rather lukewarm paper published by the FC's Forest Research following a one-year trial, described charcoal fines as a good substitute for vermiculite. David's charcoal was used in an extensive Forest Research trial, led by Tony Hutchings, into the use of charcoal in cleaning up brown-field and contaminated sites. Tony has more recently been involved in published research into using biochar in the reclamation of ground contaminated with copper and mercury, all of which show a positive result for biochar.

Taylor of Lakeland Coppice Products was asked to supply charcoal fines to an Italian research project investigating 'The biochar option to improve plant yields'. He had been using up his fines on paths, and with no fines mountain to call on, we were luckily able to supply a couple of tonnes for this project.

Finally, the yard where our fines were stored was to be sold, and they had to go. A chance mention of the charcoal fines mountain on my website led to an inquiry from one of the leading companies involved in marketing biochar, Carbon Gold. One very stressful day, thirty-three one-tonne 'dumpy sacks' (approximately 660lb/300kg dry fines weight per bag) were loaded on to a lorry and shipped away to be incorporated into their soil improver or enriched biochar.

The Evidence for Horticultural Use

Although studies considering the horticultural benefits of biochar are not universally positive, most show an improvement, some no difference, and a few are negative. The early Italian study led by Baronti noted an improvement in yields in rye grass and maize grown in pots, but found that the improvement peaked and dropped at an application rate equivalent to 60 tonnes per hectare. They concluded that this may be due to the alteration of soil pH affecting micro-nutrient availability, or even direct toxicity from the biochar (unlikely, given the source!). A more likely explanation is that at this rate of application the soil is replaced with biochar, which initially is fairly inert; authorities on biochar recommend that 20 tonnes per hectare is sufficient. An extensive review of the research undertaken in 2012 by Lori A. Biederman and W. Stanley Harpole of Iowa State University concluded:

> We find that despite variability introduced by soil and climate, the addition of biochar to soils resulted, on average, in increased aboveground productivity, crop yield, soil microbial biomass, rhizobia nodulation, plant K tissue concentration, soil phosphorus (P), soil potassium (K), total soil nitrogen (N), and total soil carbon (C) compared with control conditions. Soil pH also tended to increase, becoming less acidic, following the addition of biochar.

A literature review completed in 2017 led by Chinese soil scientist Tan Zhongxin also finds a positive message, although it is mainly concerned with the temperature at which the feedstock is produced. However, they find improvements in soil compaction, porosity, density, water content and cation exchange capacity (CEC) (*see* below), and increased microbial activity and nutrient availability.

How Does Biochar Work?

Absorbtion and Adsorption
The basic premise is that carbon in the form of charcoal has a massive surface area, and the

Lovely vegetables growing in a mix of biochar, compost, decomposed sawdust and top soil. THE COPPICE CO-OP

internal structure is able to absorb water like a sponge into its random and complex pores: one teaspoon of activated charcoal has sufficient surface area to cover a football pitch. This lends weight to the theory that charcoal produced at high temperatures >500°C is preferable for biochar, as it has a smaller pore size but larger surface area, giving greater absorption ability both of soil water and nutrients. This quality can be turned around when the biochar is oxidized or 'aged', as the charcoal shows tendencies to reduce its surface area but increase its ability to adsorb or bond with other compounds.

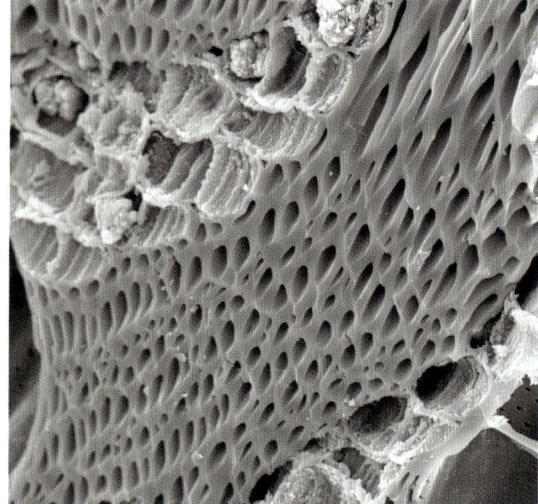

Microscopic view of charcoal. DARREN HOPKINS

Ash Die-back
(*Hymenoscyphus fraxineus*)

A few years ago the use of biochar was tested by scientists at the Bartlett Tree Research Laboratory in an attempt to improve tree resistance to ash die-back, a devastating disease that has the potential to kill 98 per cent of all ash trees planted in the UK. Biochar was added to the soil through the use of an air-spade, a tool that blasts compressed air into the soil to break up compaction. This process also mixes biochar into the existing soil, allowing it to come into contact with roots without damaging the root system. Improving tree health in an attempt to improve the trees' resilience to ash die-back disease is something that could protect the future of ash trees in the UK. Research data to date indicates that biochar could aid in this approach, as none of the biochar-amended trees were infected by ash die-back over a three-year period, while high rates of infection were recorded in non-biochar treated controls. The trial site is located in Essex.

Emma Schaffert and Dr Glynn Percival, Bartlett Tree Research Laboratory. **From:** *British Biochar News*, issue 4, May 2016.

RIGHT: *Ash die-back.*

Porosity

Charcoal is hygroscopic (water-attracting) and porous, which means that water and soluble nutrients are drawn into this complex structure. Biochar therefore reduces the leaching of nutrients from free-draining soils, and has the potential to hold the nutrients in the top strata of the soil where they can be most beneficial. Furthermore biochar's pores provide a refuge for beneficial soil fungi and bacteria. In the 'soil food web,' fungi and bacteria are at the bottom of a food chain that includes predators such as mites, protozoa, nematodes and amoebae. A protozoa can consume 50,000 bacteria a day, while bacteria can double their population every twenty minutes. If biochar is present in the soil, the predator population is reduced because their food supply is inaccessible, so bacteria and fungi can multiply. As they are both

the heart of a plant's immune system and its suppliers of nutrients from the soil, the increased population of fungi and bacteria supports increased growth and better plant health.

Compaction

For improving soil structure the biochar particle size is more important than the internal surface area as it can open up the structure and improve the (confusingly similar) porosity of the soil, leading to freer root growth and healthier plants. For this purpose biochar produced at lower temperatures is perfectly suitable.

Increased Cation Exchange Capacity (CEC)

The CEC of a soil impacts on the availability

of soil nutrients, as this is the ability of soil, in conjunction with soil microbes and enzymes, to capture nutrients from the atmosphere and make them available to enhance plant growth. In the 2017 study Zhongxin states, 'When the ratio of biochar addition is 1:100, the CEC of soil increases by 0.92 cmol/kg compared to the control, and CEC continues to increase with the addition of biochar. Returning biochar to soils improves soil CEC.' This may lead to one of the possible reasons it is not always beneficial if added to soil in a basic state, as the charcoal will become charged with the nutrients from the soil, and hold it within the structure and reduce the immediate availability to plants until their roots are able to make contact with the biochar and release the nutrients.

Soil pH

Adding biochar to soils can impact on the soil pH, mainly because the ash content of biochar can be very alkaline. Biochar produced at higher temperatures tends to have more ash and can therefore be alkaline. One school of thought favours 'weathering' the biochar to allow rainwater to wash through it, to remove ash and lower the pH. The feedstock for the biochar is also relevant – for instance, straw converted to charcoal has a very high ash content compared to that of wood.

Reduction of Plant Uptake of Contaminants

Charcoal's ability to absorb and bond with organic compounds has been shown to reduce the pollutant load of plants grown in contaminated soil. The reclamation of contaminated land by using biochar is another huge subject, with many positive research results.

Added Ingredients

For horticultural and agricultural use, biochar often incorporates additives such as nutrients and microbes which enrich charcoal fines into a ready-to-use soil improver and plant food. A healthy soil has plenty of mycorrhizae which feed off humus and break down nutrients in the soil, making them available to the plants.

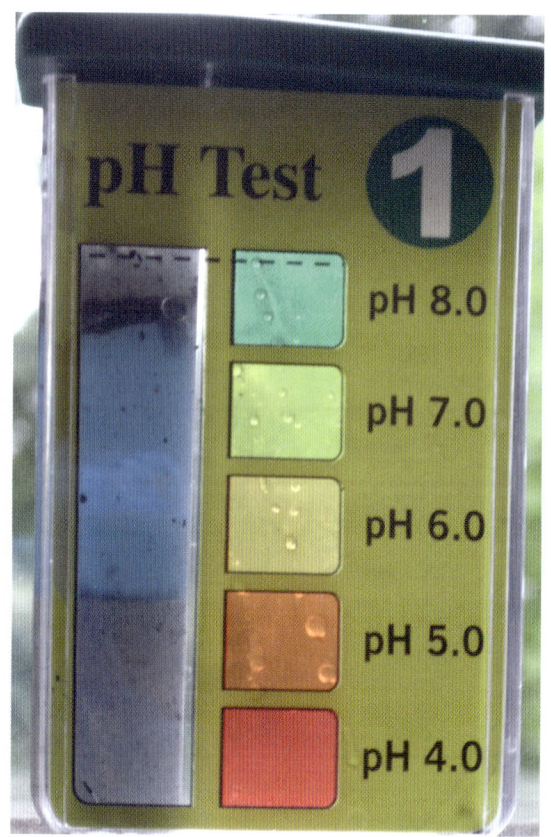

The alkaline nature of water used to quench biochar higher than eight on the scale.

Concern about the potential of raw biochar for reducing available nutrients in its inert state has led to the addition of inoculants – both microbial and fertilizers – to create a 'soil improver'. This has yet to be widely adopted, but it has a dedicated fan base and is a developing market.

The easiest way to inoculate your charcoal fines is to add them to a compost heap, but be sure to turn the heap to keep it aerated. On a commercial scale this can be done in a yard with a front-end loader. Humus can be provided by a range of waste products such as chopped woody and vegetative matter, sheep's wool, bracken or straw. Inoculant can be bought and added, and is similar to proprietary 'compost activators'.

Composting biochar with garden waste. DARREN HOPKINS

A Biochar Recipe

First, fill a 55gal (210ltr) drum with fresh water and biochar. If you are using municipal treated water, let it sit for a couple days to remove any chlorine.

Then add compost tea or worm castings and leachate to the barrel with some soil from the area where you will use the finished biochar.

For example, if you are going to apply the biochar to your fruit orchard, add some soil from around a robust healthy orchard tree. This will help charge the biochar with the ideal microbiology for your specific orchard.

Once everything is well mixed, insert a long tube such as a length of PVC pipe into the barrel, and direct air from a blower into the tube, or use a pond aerator and air stones.

Aeration supercharges the inoculant and gives the beneficial microbes a massive head start, helping them adhere to the biochar. Continue this for twelve to twenty-four hours.

Recipe taken from *Permaculture Magazine* https://www.permaculture.co.uk/articles/how-charge-biochar

CARBON SEQUESTRATION

The International Biochar Initiative (IBI) formed in 2006 soon became an influential player in the field of carbon capture and sequestration. IBI highlighted concerns about CO2 emissions and their impact on climate change, and suggested that biochar could be part of the solution. There was an inevitable backlash from groups such as Biofuelwatch, fearful that large-scale adoption of charcoal production for carbon sequestration would result in widespread monocultures of biomass grown to the detriment of natural habitats and to the loss of productive agricultural land.

There was a perception that those promoting the concept of biochar were simply seeking potential financial gains from carbon 'offsetting', and that there may be a net loss of carbon, not to mention ecological diversity, in the long term if land was brought into cultivation for biomass in the same way that biofuels have been cultivated on a grand scale. It is no coincidence that charcoal is a by-product of that biofuel industry, with flash pyrolysis the most efficient way to distil the bio oils and gases used in place of mineral oils for locomotion and power production. With flash pyrolysis, a small percentage (10–20 per cent) remains as amorphous carbon (charcoal), but this can then be buried and the carbon sequestrated.

Carbon Offsetting

Carbon credits, a key element of the Kyoto Protocol, was signed on 11 December 1997, and came into force in 2005. They take the form of a type of permit which allows a company to emit one tonne of CO2-e (carbon dioxide, or equivalent gas such as methane), and rather than producing those emissions, to trade the permit in the marketplace at a market price – thereby creating a financial incentive to not emit CO2-e. Carbon offsetting is where companies who exceed their quota of CO2-e emissions can pay for technologies or products that are low emitters, such as renewable fuels, thereby creating a financial incentive to the low carbon economy. Does biochar fit into this picture?

Carbon Capture

In 2013 the UK Biochar Research Centre at the University of Edinburgh and Heriot-Watt University joined forces with Canadian partners ICFAR at Western University, McGill University and the University of Saskatchewan to create an organization Biochar for Carbon Capture, funded by the Levershulme Trust and led by Professor Raffaella Ocone. Their remit is to work together to formalize the value of biochar as a negative emission technology.

The diagram opposite shows that the uptake of CO2 from the atmosphere by plants equals that released as the plants decay and the gas is released back into the air. So the theory that you stop the decay by pyrolysing the vegetative matter, then storing the resultant charcoal back in the ground, seems logical – but it ignores the fact that only 50 per cent of the carbon is retained in charcoal, while the rest is released in pyrolysis. We must also factor in the other potential carbon losses.

Soil Disturbance

Soil disturbance can be particularly high if the charcoal feedstock is an agricultural crop that requires cultivation to grow. Equally, howev-

Biochar's Role in the Status of Soil Carbon

In summary, field study results so far suggest that biochar is not a reliable way to increase soil carbon. It is not clear what happened to the 'lost' carbon in these different studies. Some biochar carbon might not have been stable. Some biochar may have stimulated soil microbes, which then turned existing soil organic carbon into CO2 (called 'priming'). Some may have been lost through water or wind erosion.

'Biochar: A Critical Review of Science and Policy', *Biofuelwatch*, 2011

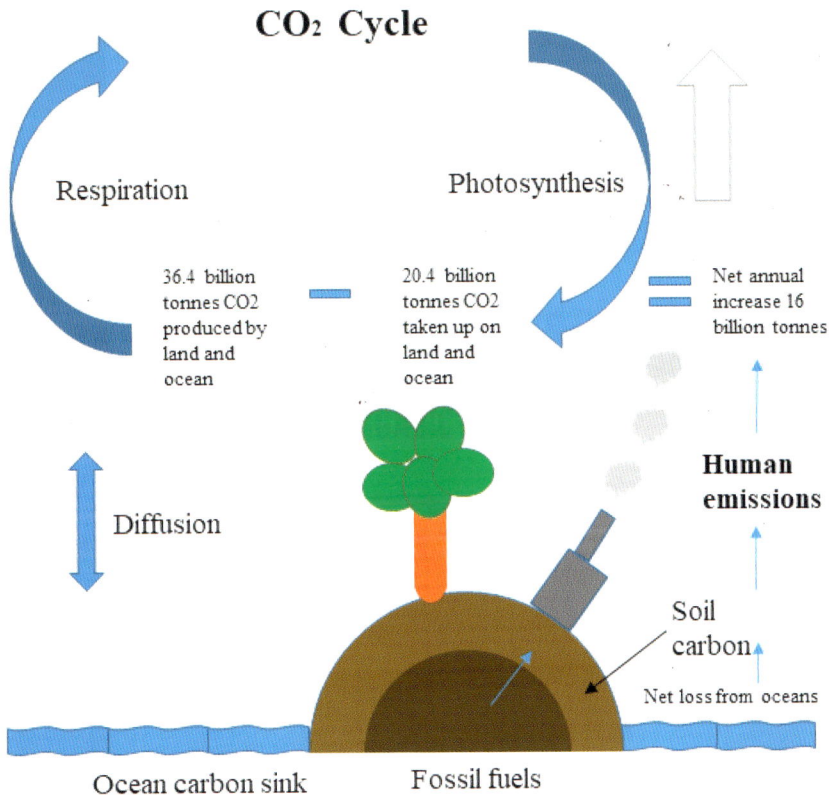

CO₂ Cycle

Respiration

Photosynthesis

36.4 billion
tonnes CO2
produced by
land and
ocean

20.4 billion
tonnes CO2
taken up on
land and
ocean

Net annual
increase 16
billion tonnes

Diffusion

**Human
emissions**

Soil
carbon

Net loss from oceans

Ocean carbon sink

Fossil fuels

Diagram of the carbon cycle.

er, there is disturbance associated with logging operations involving mature trees and heavy machinery. Disturbance allows the oxidation of soil carbon and the release of CO2. Undisturbed soils retain their carbon much better, but there is some loss over time even with carbon occurring at lower levels of the soil horizons. Just how much loss and over how long is an unsolved major area of debate, ranging from five years to 4,000 years, and any number you choose in between.

Airborne loss: There is a little loss inevitably through airborne particles (think 'dust bowl' and we can see that this can become a major factor in certain circumstances).

Waterborne loss: Some soil carbon is lost through water run-off and erosion of soil organic matter (SOM) (50 per cent carbon). Loss of SOM through flooding and extreme weath-

er events, including drought, accounts for a net loss of CO2 due to reduced soil structures stability.

Microbial Activity

As already discussed, there is a theory that adding biochar stimulates soil microbes, and that these then convert existing soil carbon to CO2, which may counteract the intended sequestration. In fact aerobic microbes will excrete CO2, but they also excrete carbon compounds, which remain in the soil and increase its fertility and carbon content as organic matter.

Kiln Technology

Methods of making charcoal vary in their efficiency. This will have an impact on climate change, particularly if we widen the debate to include the methane and carbon monoxide

Pyrolysing biochar in barrels in an open fire reduces emissions by burning off gases. DARREN HOPKINS

released in kilns that are not designed to burn these gases. We know that the impact of these pollutants on global warming is many times that of CO2 alone.

Harvesting to Processing

Transportation and timber or biomass processing also produce CO2 release. We are still very much in thrall to fossil fuels when it comes to moving biomass around: harvesting, forwarding, hauling and processing all burn CO2.

On the Plus Side?

Can all this be counteracted by plant growth? As a coppice worker my instinct is affirmative. The bleak picture of clear felling and virgin forest destruction is alien to a forest management system, which is about nurturing regrowth and renewal. Forestry is about balancing harvest-

ing with production, and anything else is deforestation. Loss of biodiversity through species change from native vegetation to exotic plantations is a threat to ecosystems and reduces the genetic pool, a policy that many would struggle to defend, at least on environmental grounds. In well managed woodlands, however, where diversity is maintained and the overall biomass is stable, surely this model can be supported – but does it add up to carbon neutrality, or better still, carbon negativity?

Efficient Processing of Waste Materials

If we step away for a moment from converting primary forest products to biochar, we can see that much could be done to process waste material to charcoal, thereby stabilizing it (for an undefined period) as carbon and sequestering it. Most mass-agricultural crops – tree crops, rice, maize and sugar cane – have waste associated with them. This waste management is a key driver for biochar, which has the potential to at least partially counteract the effect of burning fossil fuels through sequestration.

Bio-Engineering

Bio-engineering is a term which may send a chill through some, and perhaps a frisson of excitement through others, but burying the kind of quantity of charcoal required to have a significant impact on CO2 levels is certainly a major [bio-]engineering undertaking. However, if we consider the bigger picture, all human activity – from farming, fishing and forestry to urbanization – is also a massive bio-engineering project; for all the wistful desires to 're-wild' and return land to nature, the reality is that the human project has outstripped our ability to live sustainably. Perhaps, rather than sitting like frogs in our pot of ever hotter climate, we will have to act radically to reverse the core issue: we will struggle to sustain life once global temperatures increase by more than 2°C above 1900 levels.

Soil Organic Matter (SOM)

SOM is perhaps the most important benchmark of how much carbon is held in the soil rather than released in the ways described above. It is relatively easy to test to gauge whether activities are increasing or decreasing the 'carbon in the bank'. Through improved agricultural techniques, combined with the addition of biochar, where this is compatible we can measure when SOM is improving.

It has been suggested that we should set aside carbon trading because it can lead to unintended negative consequences, and focus on a 'carrot and stick' approach to soil carbon: if you increase it, you get credit – if you decrease it, you would be fined. Craig Sams, in a presentation to the Harmony in Food and Farming confer-

ence in Llandovery, Wales, in July 2017 (unpublished) sets out the concept of carbon pricing. He calculates that an organic farm might make £100 a hectare per year by increasing SOM and receiving 'interest' on the carbon in the bank, whereas a factory farm with practices that reduce SOM would pay £100 per hectare per year – quite an incentive to address this issue.

CONCLUSION

Biochar has a beneficial effect on plant growth and can be used to good effect in agricultural and forestry scenarios. All biochar is not equal, and attention must be paid to the variables in order to match the suitability of the biochar to each scenario. Variables can be summarized as:

- Feedstock: hardwood, softwood, short rotation coppice, straw, husks, nutshell, sewage sludge etc.
- Pyrolysis temperature: slow 200–400°C, medium 400–600°C, fast (or flash) 600–1000°C
- Particle size: 0 to 25mm
- Additives: nutrients, microbes, enzymes, humus

Harnessing these benefits, and using them in particular to enhance the productivity of subsistence agriculture, where small-scale recycling of biomass as charcoal can be used to great effect, is a positive development. Far from being a panacea, there are many barriers to adoption of these techniques, including a key one of land ownership and security of tenure. To look to the future beyond a quick cash-crop takes commitment, foresight and long-term planning. Did the Amazon people know they were creating something with such long-term benefits? Possibly not, but one benefit of the longevity of their soil technology is that we can still learn an important lesson from them, which includes using biochar to increase fertility.

Chapter 10
A to Z of Charcoal Products

Over the centuries, charcoal in its different forms has been used in a remarkable number of products. In recent years the large charcoal manufacturers have developed an ever-increasing variety of charcoal-based products. However, although most contemporary small-scale burners have been unable or unwilling to tap into many of these markets, there are opportunities for charcoal makers to access this diversity, given the desire for locally sourced and specialist ingredients and products. The various niche markets include barbecue charcoal, biochar, and charcoal fines in various particle sizes, from coarse to granular or ground into powder.

This final chapter considers the possibilities for using charcoal across the full range of products, which are set out in the form of an A to Z. Hopefully this chapter, and the book as a whole, will help stimulate further new creative ways that charcoal burners can develop, and which will help to market their products.

Activated charcoal: Many medical and filtration uses for charcoal require what is termed 'activated' charcoal. In this process either steam or chemicals are used to erode the extensive pores on the surface of the charcoal particles, which increases the surface area of the material,

Willow drawing charcoal.

and thus the capacity for the charcoal to adsorb impurities. Activation is only safely undertaken in industrial-/laboratory-based contexts, as there are many hazards involved in handling the chemicals, and steam activation requires extremely high temperatures.

Steam activation: Steam activation involves passing supercharged steam over charcoal at temperatures of 926°C to 982°C, which are hard to achieve in a non-industrial situation. The steam, the heat and the absence of oxygen create the conditions for the conversion of a proportion of carbon to hydrogen gas and carbon monoxide, which is vented off. One third of the original weight is lost when converting charcoal to activated charcoal. The effect on the charcoal structure is to open the multiple surface 'pores', which greatly increases the surface area. The resulting activated charcoal is then washed to remove ash and residual wood vinegars.

Chemical Activation: The alternative is the chemical activation process, which can be done at temperatures of between 450°C and 900°C, using phosphoric acid, which erodes the pore surface of the charcoal, with the same effect of increasing the surface area. The resulting activated charcoal will need extremely careful washing to remove all traces of phosphoric acid, especially if the charcoal is intended for human consumption.

Agricultural use: Charcoal is known for its ability to neutralize odour, making it ideal to incorporate into stock bedding and slurry storage systems.

Animal feed: Charcoal has long been fed to animals, particularly to those ruminants with more than one stomach. The charcoal absorbs digestive tract enzymes, and can harbour a concentration of bacterial exo-enzymes, which improve the efficiency of the ruminants' stomachs to digest food and access nutrients. Charcoal also has the ability to bond with some toxins, which are then excreted, so the charcoal in feed supplements may also support the immune system.

Granular charcoal as horse feed supplement.

Various proprietary food supplements for agricultural animals and pets contain charcoal, including Fine Fettle Feed's 'Happy Tummy' for horses (and other mammals) and the Granular Animal Charcoal Feed Supplement produced by Dorset Charcoal. These companies claim that animals may benefit from a charcoal feed supplement, because charcoal binds with toxins, enhancing overall health.

Artist charcoal: Traditionally artist charcoal is made from willow sticks, though in fact any wood can be used (each different type of wood produces charcoal with a different texture). Willow charcoal's softness allows a smudged or blurred effect to be produced. Charcoal made with hazel is best for a harder line.

Small-scale artist charcoal may be made in a tin in a kiln, barrel or retort. Allow an air-hole for the gases to escape, and bury the tin in the bulk of the timber stacked in the kiln.

Feeding Biochar Survey in Dairy Cattle

In a study on biochar in agriculture by Achim Gerlach and Hans-Peter Schmidt, twenty-one farm managers were recruited from farms with an average herd of 150 cattle. They were asked to feed the animals 200–400g of biochar per cow per day, and observe changes over four weeks. Some biochar was also mixed with sauerkraut brine and fed to cattle with additional health problems, with beneficial effect. They reported the following changes in the cattle, within one to four weeks after the introduction of biochar in the feed:

• Generally improved health and appearance
• Improved vitality
• Improved udder health
• Decreased cell counts in the milk (interrupting the administration of biochar leads to higher cell counts and a drop in performance)
• Minimization of hoof problems
• Improved post-partum health
• Reduced diarrhoea within one to two days, faeces subsequently generally more solid
• Decline in the mortality rate
• Increase in milk protein and/or fat
• Marked improvement of slurry viscosity, with less stirring needed and less scum on the surface
• Improvement in the smell of the slurry
• Also an increase in ammonium nitrogen and a reduction in nitrate and nitrite in slurry produced by the cows

Gerlach and Schmidt noted that high-grade biochar (produced to the European biochar standard) for animal feed still gave good results, and it was not found necessary to use the medical-grade activated charcoal, even though the biochar is about a third less absorbent.

From: Gerlach and Schmidt, Journal for Terroir-wine and Biodiversity, 2008, ISSN 1663–0521

Make your own Artist Charcoal on a Camping Stove

• Take a metal tin (many tins nowadays have a plastic element) – golden syrup tins are suitable, and have a well sealed lid
• Puncture two or three holes in the lid with a hammer and nail
• Cut to length some peeled willow sticks – or a log split into square sticks – to fit into the tin, so the lid will still shut
• Put the lid on and place the tin on the stove, outdoors
• The wood will heat up and start to give off gas, which you can light and watch it burn
• When the flame goes out, the gases are spent, so switch off the stove and allow the tin to cool

Charcoal drawing by Andrea Pentecost.
ANDREA PENTECOST

Willow charcoal made in a tin in the charcoal retort.

Building: Charcoal has several uses in the building industry: insulation, humidity regulation, soil decontamination, and as protection against electromagnetic radiation.

Insulation: Charcoal's porous nature enables it to trap both air and liquids, making it an efficient insulation material. Experiments have been conducted to establish the effectiveness of biochar as a building material, by mixing biochar into plaster for rendering the inside and outside of buildings. The biochar is mixed with lime mortar or cement (as much as 80 per cent charcoal), which can be applied up to 8in (20cm) thick.

Humidity regulation: As charcoal adsorbs moisture from the atmosphere it can be used both to help regulate the building's internal humidity and to neutralize smells. For internal plastering a mix of 50 per cent charcoal, 30 per cent sand and 20 per cent clay can be applied, using coarse char up to 10in (25cm) in the first layer, and finishing with a finely ground char to give a smooth finish. It is also used as an under-floor layer to regulate moisture by filling the cavity between the floor joists. The layer traps moisture in the damp seasons, releasing humidity when the surrounding conditions are very dry.

Soil decontamination: Charcoal is proven to remove many heavy metals from the soil and has been successful, for example, in treating ground contaminated with cadmium, lead and copper.

Protection against electromagnetic radiation: Shielding paint is available, which claims to cut out electromagnetic radiation, using 'only carbon'. Electromagnetic radiation affects some people badly, and has security implications for electronic data, so it can be very important to shield people and equipment from these rays.

Burnishing: Finely ground charcoal is used for burnishing metal, as it brings up the brightness without scratching the surface. There is an example of this in a lovely video from the Victoria & Albert Museum of Korean lacquered boxes, where the mother-of-pearl shell is cut in intricate patterns and applied to the lacquered box with glue. The charcoal paste is then rubbed in to remove glue residue, and to polish the shell. Paint and more lacquer is then applied to a perfect finish. See www.youtube.com 'How a traditional Korean inlaid lacquer box is made'.

Cosmetics: 'Charcoal fines are good for your complexion!' A range of soaps and cleansers with charcoal as an ingredient is available, and a scan of the cosmetic aisle suggests that good marketing can always persuade someone to buy an attractive product. The quality control for products containing charcoal is stringent, to avoid contamination from soil, ash or sand. The charcoal is ground down to a very fine powder before being added to the toiletry product.

Charboard: Ground charcoal can be mixed with wood pulp for making paper and cardboard. This could have the benefit of reducing and recycling waste from paper-making factories into charcoal, which can be added as a filler to packaging products. (Draper and Schmidt, 2014.)

Soap made with charcoal fines.

Dog biscuits: Charcoal is an ingredient in dog biscuits sold as a mix in some major brands. It is also possible to make charcoal dog biscuits at home.

Electric light bulb: Early technology for electric lights included a carbonized bamboo filament. The American inventor Thomas Edison made a light bulb in the 1880s using Japanese Madaka bamboo for the filament.

Filtration: The most common use for charcoal in filtration is activated charcoal in domestic water filters. Other, less well known specialist filtration uses are mentioned below.

Decaffeination: The Swiss Water Method of decaffeination uses activated charcoal filters – it has been suggested that it should be called the Swiss Charcoal Process. The green coffee beans are soaked in very hot water to dissolve the caffeine. The water is then passed through an activated charcoal filter which absorbs 99.9 per cent of the caffeine, while retaining the flavour. The green coffee beans are returned to the solution to reabsorb the flavour. This process is

---

Make Your Own Dog Biscuits!

1lb flour, wholemeal or a mix of wholemeal and white
8oz oatmeal or porage oats
2oz food-grade activated charcoal
2tsp dried yeast

Mix the dry ingredients

Add 2tbsps olive or sunflower oil

Mix to a dough with about 10fl oz of warm chicken stock, and knead until elastic.
Roll out to a 1/4in thick, and cut into biscuit shapes of your choice.

Leave to rise for 60min, then bake at 150°C for 50–60 minutes, turning off the heat to allow the biscuits to cool in the oven.

Store in an airtight tin.

Decaffeinated coffee using the Swiss Water Method.

considered one of the best ways of decaffeinating coffee.

Gold extraction: Gold is dissolved in a cyanide solution, then passed through a series of carbon filters, which 'collect' the minute gold particles – up to 9–18lb (4–8kg) of gold per tonne of charcoal. The system can be used to extract any metals capable of being held in a solution and then filtered out through charcoal.

Filtration through non-activated charcoal: Non-activated charcoal – essentially this is biochar – also has a role in filtration. An improvised pond filter can be made by placing a sack of charcoal fines in the pond – but only use 'weathered' or cured fines, which have a fairly neutral pH.

Fireworks: *See* Gunpowder (*see* Chapter 5). A note of caution: you do need a licence to manufacture fireworks or any explosive material.

Fish farming: Activated charcoal filters are used in aquaculture to purify the water. Fish farms use various chemicals as disinfectants, medicines and insecticides to treat diseases and parasite attacks. Through absorption, charcoal filters can prevent the chemicals from building up to potentially toxic levels.

Foodstuffs: 'Charcoal biscuits are good for the digestion!': a favourite advertising slogan dating back to the nineteenth century when charcoal biscuits were widely available as an aid to digestion. The modern trend for seeking the most healthy diets and super-foods has not ignored the humble charcoal biscuit. Often sold as a gourmet cheese cracker, they can be made with activated willow charcoal added to a simple biscuit dough. There are, however, concerns that ingestion of too much food-grade charcoal can lead to the unwanted outcome of good nutrients not being absorbed – but this is unlikely if only eating the occasional charcoal biscuit.

Food supplements: Food-grade activated charcoal has been marketed as a food supplement. Though charcoal has no food value it is often taken to aid detoxification programmes, with the aim of supporting beneficial gut flora and digestions as the charcoal binds with, and thus removes toxins. Care, however, must be taken not to overdo charcoal ingestion, especially those taking a medication that could bind to the charcoal, rather than be absorbed by the body.

Charcoal Biscuit Recipe

1lb flour
3oz butter
3oz sugar
1½oz activated charcoal
1 egg, beaten
Milk to mix

Rub the butter into the flour, mix in the charcoal powder and the sugar. Make a well and pour in the egg, stir in the flour and mix to a biscuit dough, adding a little milk if needed. Roll out fairly thin, cut out to any desired shape, wash with water, and bake in a medium oven until well done.

Buckminsterfullerene (C_{60})

Buckminsterfullerene (C_{60}), or Fullerine C60 as it is also known (described in Chapter 1), when compressed, is harder than diamond – and in its uncompressed form it has been tested as a potential food supplement. There was quite a stir in 2012 when a trial on rats designed to test if Fullerine C60 was toxic revealed that the rats had a 90 per cent increase in their life expectancy when fed a 1.7mg per kilogram bodyweight dose of Buckminsterfullerene. The most likely reason for this is that it acts as an anti-oxidant and reduces free radicals, which can be harmful to cells in the human body. Fullerene is now marketed as an ingredient in an anti-ageing cream.

Gourmet foods: There is a definite trend in the past few years for charcoal to appear in some form, whether as an ingredient or as a flavouring in smoked food, on the menus of the top restaurants in Britain and around the world. Food critics wax lyrical about the smoky 'umami' flavour and the visual impact of an all-black plate of food. The photograph opposite (top) pictures an example of gourmet charcoal-infused food, available from a high street supermarket: the 'Heston Charcoal Bagel with Tea Smoked Salmon' (which was rather tasty!)

Hair shampoo: The claims for this product are that the charcoal will absorb excess oil, dirt and impurities, allowing them to be washed away.

Hand warmers: These warmers have charcoal inserts that burn slowly and give off heat. They are available in a number of makes from 'outdoor' shops.

Incense burning: Incense sticks are made with a mix that includes sawdust and often charcoal. You can make your own incense to burn using a finely powdered charcoal, preferably Binochan or Japanese charcoal. In the past, charcoal was made from alder buckthorn (*Frangula alnus*), always the species of choice for gunpowder makers; however, these days it should not be used, as it is quite rare – though it can still be found in limestone areas of Britain. Alder (*Alnus glabra*) charcoal would be a good substitute.

Inks and pigments: The 'Old Masters' used a paint called bone black or ivory black, which was made by charring bones (or ivory, for preference) and grinding it with linseed oil. Charcoal is still used as a pigment in paints and drawing inks. You can make your own watercolour paint by mixing a combination of finely ground charcoal powder and gum arabic.

Japanese charcoal or Binchotan: Binchotan, or white charcoal, is produced in Japan in stone and clay kilns from Japanese holm oak or the ubame oak (*Quercus Phillyraeoides*). The coppiced branches of the oak are stacked in the kiln and the air inlets restricted to allow it to burn slowly at low temperatures for three days. On the fourth day the air is slowly increased until the wood in the kiln reaches 900°C–1200°C. At the right moment the wood is raked out of the kilns with long-handled rakes and covered

Heston from Waitrose: 'Charcoal Bagel with Tea Smoked Salmon'.

Beauty Formulas Shampoo with Activated Charcoal.

in ash and red clay to exclude the oxygen and allow the wood to cool. When cold, the charcoal sticks ring like metal when struck, and the cut ends are glassy. This charcoal is an essential part of the ancient tea ceremony of Japan.

Kitchen uses: Apart from the obvious use in the kitchen of cooking on charcoal, charcoal can be put in the fridge with vegetables and fruit to absorb the ethylene produced as part of ripening, to prolong freshness.

Lime and cement manufacture: Charcoal fines are mixed with lime in a kiln then fired, a process still used in Somalia. This type of lime cement was once widespread in Britain, and old lime kilns are still found in many places.

Medical use

Poisoning: Charcoal is routinely used where patients have swallowed poison. It should be administered as powdered activated charcoal in a Sorbitol solution within an hour, if at all possible. The charcoal works by absorbing the poison and allowing it to pass safely through the body. It does not work, however, with poisons such as cyanide and alcohol.

Charcoal biscuits for digestion: Charcoal has been added to foodstuffs for centuries.

Carbon as a homoeopathic and herbal remedy: Carbo Veg is a remedy widely used in homoeopathy and herbal medicine. It is indicated where there is a picture of great weakness, breathlessness or faintness. It is also useful in indigestion with flatulence and bloating.

As a binder to administer drugs: Because of charcoal's ability to bind with other compounds, it can be used to bind with drugs for medical purposes to take them into the body.

Nanotechnology: Bamboo charcoal is ground to a fine powder and embedded into rayon to make fabric. This powder is so fine the particles are a mere 1–100nm long: nanoparticle size. Rayon has various textile uses, among them the inserts in babies' disposable nappies, harnessing the properties of charcoal of odour neutralizing and absorbency. There are some concerns about nanotechnology, and as yet the lack of research into this relatively new technology means the risks are theoretical and untested.

Nano particles
Nano particles are 1–100 nanometres in size. A nanometre is one billionth of a metre or 0.000001 of a millimetre.

Lime kiln.

Charcoal path.

Organic agriculture and gardening:
Charcoal fines (as biochar) can be used as a natural soil improver, to help build up the soil structure, condition and fertility. Charcoal has a place in organic horticulture and agriculture.

Paths: Because charcoal does not decay or break down readily it can be a useful medium for paths. It is best avoided in garden situations where it will be walked into the house, but is effective in allotment, farm and woodland settings.

Pellets: Activated charcoal is sold as pellets for adding to aquariums and for ponds, and for refillable water filters and suchlike.

Quirky items: See the charcoal pine cones on page 166.

Refrigeration: Charcoal can be used to make a screen using two layers of wire netting with the charcoal in between. The charcoal is then wetted with cold water, and any breeze that passes through the screen is cooled, giving relief in very hot climates.

Rhododendron as a feedstock for charcoal: In the 1990s, there was great concern that rhododendron was toxic to humans and could affect people who ate food cooked on rhododendron charcoal. The symptoms of rhododendron poisoning in cattle and sheep are vomiting and weakness, which can lead to death. It was noted, however, that it would be necessary to ingest a great deal of sappy growth to feel any effects of these toxins. Although there may be some risk from breathing in rhododendron smoke during the making of charcoal, there would be so little of these compounds left following distillation that the risk to the public would be negligible.

This leaves the charcoal burner with a dilemma as to how much they want to handle and breathe the smoke from rhododendron burns. It would seem that a sensible approach would be to avoid breathing smoke when making charcoal whatever the feedstock, and never ingest rhododendron leaves. As a postscript to this story, there is a move to cut back widespread unwel-

come invasive *Rhododendron ponticum*, and turn the woody mass into biochar. This can then be put back into the soil to ameliorate the poisoning effect of rhododendron rhyzomes, which deter any other species from colonizing the soil. (Craig Sams pers.comm)

Semi-conductors: Semi-conductors are found in all modern electronics – unfortunately, although it is beyond the scope of this author to explain how it works, it seems that charcoal may become the next 'big thing' in the world of electronics.

Textiles: Charcoal's insulating and odour-control properties make it an obvious choice for incorporating into clothing, specifically sportswear and shoe insoles.

Teeth cleaning: Toothpaste is now available on the market which is made using activated charcoal made from either coconut shell or Binchotan charcoal from Japan. The charcoal is ground into a very fine powder and is abrasive but not damaging (unlike salt-based products). The products claim to remove stains from teeth, and may have a flavouring such as peppermint, or be flavourless, such as the one from Live Co Co.

A toothbrush with charcoal-infused bristles is also available. Claims for this include the following: 'Charcoal has been long known as one of nature's best absorbing and cleaning agents. It is a natural antibacterial, antifungal and antiviral agent. It detoxifies and deodorizes your mouth, balancing PH levels while absorbing odour.'

Tattoos: Only the very dedicated would go to the length illustrated opposite to celebrate charcoal burning.

Universal cooker hood filter: Carbon filters are often used in cooker hoods to filter out cooking smells.

Vehicle emissions: Filtration of fumes and particulates.

Vinegar: Wood vinegar – a mix of various organic substances, including a high proportion of acetic acid – is a by-product of char-

ABOVE: *Invasive* Rhododendron ponticum *crowding out native vegetation.*

RIGHT: *Tattoo of charcoal burning.*
NICK HARRIS

Charcoaled pinecones.
JEAN HAYMAN

coal making, and has many uses. An unusual one is as a bird deterrent. Bottles of wood vinegar are suspended in the trees with a few holes in the top to allow the fumes to come out, thereby deterring birds from roosting in the trees. It has been used to good effect in Japanese cities, where massive starling flocks present a health hazard and noise nuisance. It does not harm the birds, which hopefully find somewhere else to roost.

Wood vinegar is used on crops, particularly in Japan, as a pest repellent and a fungicide, and when mixed with biochar, as a soil improver.

Water filters: Many domestic and industrial water filtration systems use activated charcoal. It is a very effective medium for removing discoloration or smells where there is rotting vegetative matter in the water. One particular use is to filter distillery waste water, which has been described as 'highly coloured and acidic with a strong and objectionable odour' (Singh 2017).

Wind chimes: White charcoal or Binchotan from Japan and other Far Eastern countries has a metallic ring, and when hung up on strings, makes a beautiful-sounding wind chime.

Xylem in charcoal: The main cells in wood are phloem and xylem. Phloem is described as the living transport system of a tree, and xylem the non-living, though they also carry and store water and minerals. 'When a xylem cell reaches maturity it is reinforced with lignin and then dies' (Reitz 2012). This is what charcoal is made of.

Yield improvement with biochar: A study in Nepal found a fourfold increase in productivity of pumpkin plants using biochar as a carrier for cow urine in compost. In Ghana, improvements in maize yields were reported ranging from 200–250 per cent, with applications of a 50/50 biochar compost mix. A study in Germany looked at the benefits of biochar on grape-vine growth. The study compared

Picture of tooth whitener.

a control (no compost or biochar and a light sandy soil) with vines grown in compost alone, and another with vines grown in compost and biochar. This did not show a significant difference between the two growing media, but the benefits may need time to show up as many studies report a drastic reduction in leaching of nutrients in soils augmented with biochar. (Niggli and Schmidt 2012.)

Zimbabwean biochar: The African Biochar Partnership was launched during an international workshop at the World Agroforestry Centre, Nairobi, in early March 2016. It provides support for projects such as The Biochar Plus project, which is co-funded by the EU and the African Caribbean and Pacific States Co-operation Programme in Science and Technology (ACP Science and Technology), involving partners from Cape Verde, Ethiopia, Ghana, Sierra Leone, Togo, Zimbabwe and Italy. The Zimbabwean involvement is through the Binduri University of Science Education. The project is led by Professor Alessandro Perressotti of the University of Undine in Italy.

The Partnership intends to provide knowledge-sharing and agreement of standardized methods of producing biochar, and will cover a broad perspective including agriculture, energy, waste management and health.

Appendix
Comparative Table of Charcoal Production

Type of system	Initial cost (£)	Main-tenance costs	Timber required per burn (kg)	Hours per burn	Conver-sion rate	Charcoal made per burn (kg)	Charcoal made per week (kg)	Notes
Exeter Retort	14,000	High	750	8	01:04	150	900	Can run more than one, sits on a trailer
Pressvess Retort (twin chamber)	20,000	High	1,400	Contin-uous	01:04	350	2,450	Needs to be kept hot
Pressvess Retort (single chamber)	16,750	High	400	6	01:04	100	600	
Rocket Retort	3,000	Low	100	4 to 6	01:04	25	150	Can run more than one
Oxford Retort	100,000	High	2,500	Contin-uous	01:04	600	7,000	Needs to be kept hot
Viper Mobile Kiln	—	High	1,000	Contin-uous	01:04	250	5,000	Sits on a trailer
Ring Kiln (8ft)	1,300–1,500	Low	2,000	10 to 28	01:08	250	500	Can run more than one
Ring Kiln (5ft)	1,100	Low	1,000	6 to 12	01:08	125	375	Can run more than one
Kontiki Kiln	1,200	Low	2,400	8	01:08	300	1,800	Can run more than one
Barrel Burn	10	Low	80	8	01:08	10	60	Can run up to 10 with one person
Earth Burn	Nil	Nil	2,000	48	01:08	250	750	Can run more than one

Glossary

ancient woodland A woodland that has existed continuously since at least 1600AD and therefore possibly since pre-history.

barking Peeling bark from a tree, normally oak, for use in tanning.

bast The inner layer of the bark of elm or small-leaved lime, used for chair seating.

bavin A bundle of brushwood or firewood

beetle Large-headed and often long-handled mallet used for jobs that require a lot of momentum behind the hitting action.

bender Temporary shelter made using bent poles covered by polythene or canvas sheeting.

bill hook Ubiquitous hand tool with a long blade often ending in a hook, used for coppicing, dressing out and riving hazel.

billet A short length of wood, varying from about 8 to 36in (20 to 90cm), sometimes split.

binder (also sway) Long hazel rod used to fix thatch to a roof frame.

biochar Small particle charcoal that is incorporated into the ground.

bolt Round log blank before being used to make shakes.

brash (also brish or brushwood) The small twiggy branches resulting from dressing out the tops and side branches from coppice or standards.

brands Wood that has only been partially converted to charcoal.

brazier Receptacle for burning a fire off the ground.

brown ends Wood that has only been partially converted to charcoal.

butt The lowest portion of a trunk, stem or pole.

char or charking Turn wood to charcoal.

clamp The earth kiln.

cleft To split a segment of wood from a round pole; see also rive.

coalwood Small-diameter wood for charcoal making.

cooler Also as rabil, a tool for drawing the earth off the clamp when the charcoal is made; a flat board at right angles to the long handle.

coppice Underwood trees that are cut close to ground level every few years to allow multiple stems to grow again from the stool.

coppice-with-standards System of coppice management with scattered, single-stemmed trees such as oak or ash (*see* standard).

coppicing cycle The number of years between cutting of the coppice – see rotation.

cord A volume measurement of timber 4 × 4 × 8ft.

cowrake or carrack Like a cooler but also more of a rake.

crown The living branches of a tree above the main stem.

faggot Bundle of twigs and small brash tied tightly for use as a firelighter.

firgun stick A pointed stick for making a hole to pour water in when quenching the burn.

flaying Sprinkling the earth burn with water.

greenwood Freshly felled living wood, still retaining its sap.

hardwood Any broadleaved tree, irrespective of the actual hardness of the wood; see softwood.

hearth The site where the earth burn is built and burnt.

heartwood The inner wood of large branches and trunks, which no longer carries sap.

hurdle A windbreak for protecting the earth burn.

mallet A wooden hammer.

maul Often known as a sledge-hammer, but can be a broad-headed mallet.

moisture content The amount of water remaining in a piece of wood, measured with a moisture meter.

motty peg A piece of wood that is in the centre of the earth burn and removed.

mould The earth used to cover the earth burn; may be sieved and reused.

overstood Coppice that is still standing

beyond its normal rotation.

pitstead The hearth where the earth burn is built.

platform As for hearth.

pleachers The cut stems of a hedge when it is laid to regenerate it.

prog A stout forked pole used for pushing and levering trees during felling, or for turning the remains of a fire.

rabil, rubler, rebel or rauber A tool such as a cooler for drawing the earth off the clamp when the charcoal is made.

rive To split or cleave a piece of round wood; more often used to describe the splitting of smaller pieces.

rod Small, flexible underwood stem of less than 2in (50mm) diameter.

rotation Length of time between the cutting of a coppice coupe; see coppice cycle.

roundwood Wood of small diameter often used for fencing stakes.

sacrificial wood Wood burnt to ash to get the kiln up to temperature.

sammel A sandy subsoil that is used to cover the earth burn; can be sieved and reused.

sapwood Wood that carries the sap within a tree stem; this may be all the wood in a young stem, or the outermost layer in an older, larger trunk or branch.

shanklin The thicker logs that are placed in the centre of the burn where the temperature is higher.

sod drag A long-handled fork that has tines at a right angle to the handle so it will pierce the turves and drag them off the earth burn.

softwood The timber of a coniferous tree, irrespective of the hardness of the timber.

stail A long handle.

standard A single-stemmed tree, never coppiced or pollarded; any tree not grown from a coppice stump.

stangs A frame with four handles used for carrying water barrels.

stocker A right-angle spade such as a large adze, which is used for clearing ground.

sway (also binder) Long hazel rod used to fix thatch to a roof frame.

swill (or spelk) Basket made from woven, boiled oak sapwood strips, now only made in the Lake District.

timber Tree trunk suitable for making beams or for sawing into planks, normally derived from standards.

tine The tooth of a rake or fork.

trunchion A length of wood.

turf or turves Grass- or heather-covered earth cut into squares or oblongs for covering the clamp.

underwood Coppiced shrub layer growing under standard or timber trees.

wood The part of the stem inside the cambium which supports the tree, carries water to the crown, and stores reserves of food over the winter period.

Bibliography

Armstrong, L. *Woodcolliers and Charcoal Burning* (Coach Publishing House Ltd, 1978)

Baronti, Silvia, Giorgio Alberti, Gemini Delle Vedove, et al. 'The Biochar Option to Improve Plant Yields: First Results From Some Field and Pot Experiments in Italy' Italian Journal of Agronomy (online), 5.1 pp. 3–12 (2010)

Biederman, L. A. and Harpole, W. S. 'Biochar and its effects on plant productivity and nutrient cycling: a meta-analysis' GCB Bioenergy 5: pp. 202–214 (2013)

Bramwell, M. (ed) *The International Book of Wood* (Mitchell Beazley Publishers Ltd, London, 1976)

Bond, T. C. and Sun, H. 'Can Reducing Black Carbon Emissions Counteract Global Warming?' Environ. Sci. Technol., 39 (16), pp. 5921–5926 (2006)

Cervenka, J. 'Harder than diamond, stronger than steel, super conductor … graphene's unreal', The Conversation, March 18 (2012) http://theconversation.com/harder-than-diamond-stronger-than-steel-super-conductor-graphenes-unreal-5123

Morris, John, 'Chilterns Charcoal Workshop: Conference proceedings' (1998) http://www.chilternsaonb.org/caring/woodlands_project.html

Donald, M. B. *Elizabethan Copper* (Red Earth, 1994)

Draper, K., and Schmidt, H. P. 'Biochar Paper – elevating biochar from novelty to ubiquity' Biochar Journal, 26 August (2014) www.biochar-journal.org/en/ct/15 (Accessed: 28.08.2017)

Edlin, H. E. *Woodland Crafts in Britain* (David and Charles, 1979)

EBC. 'European Biochar Certificate Guidelines for a Sustainable Production of Biochar' European Biochar Foundation (EBC), Arbaz, Switzerland (2012), version 6.3E, 14 August 2017 http://www.european-biochar.org/en/download

Fisher, W. R. Schlich's *Manual of Forestry Vol V. Forest Utilisation* (Bradbury, Agnew and Co. Ltd, London, 1908)

Evelyn, *John Sylva; or, A discourse of forest-trees, and the propagation of timber in His Majesties dominions. As it was deliver'd in the Royal Society, the XVth of October, CIƆIƆCLXII .(1664 edition)* (reprinted Reink Books, 2017)

Frondel, C., and Marvin, U.B. Lonsdaleite, a new hexagonal polymorph of diamond Nature, 214 (5088): pp. 587–589 (1967)

Gerlach, A., and Schmidt, H-P 'The use of biochar in cattle farming' Biochar Journal, Arbaz, Switzerland, 1 August (2014) www.biochar-journal.org/en/ct/9 (Accessed: 31.08.2017)

Gordon, A. 'Richard Watson, 1737–1816' Dictionary of National Biography, 1885–1900, Volume 60 https://en.wikisource.org/wiki/Dictionary_of_National_Biography,_1885-1900/Vol_60_Watson_-_Whewell

Gray, M., Johnson, M. G., Dragila, M. I., and Kleber, M. 'Water uptake in biochars: The roles of porosity and hydrophobicity' Biomass and Bioenergy, 61, pp. 196–205, (February 2014)

Gurwick, N.P., Moore, L.A., Kelly, C. and Elias, P. 'A Systematic Review of Biochar Research, with a Focus on Its Stability in situ and Its Promise as a Climate Mitigation Strategy.' PLoS ONE 8(9) pp. 1–9 (2013)

Harris, P. 'On charcoal', Interdisciplinary Science Reviews, 24:4, pp. 301–306 (1999)

Hamatui, N., Naidoo, R. N. and Kgabi, N. 'The respiratory health effects of exposure to charcoal dust' International Journal of Occupational and Environmental Health, 22:3, pp. 240–248 (2016)

Herd, A. R. C. 'Exploring the socio-economic role of Charcoal and the potential for sustainable Production in the Chicale Regulado, Mozambique', unpublished dissertation: University of Edinburgh (2007)

Kelley, D. W. Charcoal and Charcoal Burning (Shire Album, 2002)

Hollingdale, A. C., Krishnan, R., and Robinson, A. P. *Charcoal Production, A Handbook* (Natural Resources Institute, 1999)

Howkins, C., *Trees, Herbs and Charcoal Burners* (Chris Howkins, 1994)

Karasavvas, T. 'Ötzi's Ancient Axe is from Tuscany, Giving Firm Evidence of Neolithic Travel and Trade' http://www.ancient-origins.net (Accessed 11 July 2017)

Klar, M. *The Technology of wood distillation* (Chapman and Hall, 2nd edition p. 8, 1925)

Lehmann, J., and Joseph, S. *Biochar for Environmental Management: Science and Technology* (Earthscan, 2009)

Lewis, H. 'Charcoal and wool: The role of woodland in the worsted revolution of West Yorkshire' (Conference Paper: University of Bradford, 2017)

Linnard, W. 'Sooty Bands from Tents of Turf. Woodcolliers and Charcoal Burning in Wales', Folk Life, 25:1, pp. 47–73 (1986)

Living Heritage, Parliament and the East India Company http://www.parliament.uk/about/living-heritage/evolutionofparliament/legislativescrutiny/parliament-and-empire/parliament-and-the-american-colonies-before-1765/parliament-and-the-east-india-company/ (Accessed 31 August 2017)

Marris, E. 'Putting the carbon back: Black is the new green' Nature, 442, pp. 624–626 (10 August 2006)

MERL. Charcoal Burning (blog) https://blogs.reading.ac.uk/sense-of-place/tag/charcoal-burning/ (Accessed 31 August 2017)

Miles, T., and West, T., *Quest for Barbel* (The Crowood Press, 1991)

Morrison, R. *Carbon Organic and Hydrocarbon Remedies in Homeopathy* (Hahnemann Clinic Publishing, 2006)

Niggli, C., and Schmidt, H.P. 'Biochar in European Viticulture', Ithaka Journal 1, pp. 13–25 (2012)

Oaks, R., and Mills, E. *Coppicing and Coppice Crafts – a comprehensive guide* (Crowood, 2010)

Rackham, O. *Trees and Woods in the British Landscape* (Phoenix Giant Paperback, 1976)

Rackham, O. *Woodlands* (Collins, 2006)

Reitz, E., and Shackley, M. *Environmental Archaeology* (Springer, 2012)

Robinson, N., Evershed, R. P., Higgs, W. J., Jerman, K., and Eglinton, G. 'Proof of a pine wood origin for pitch from Tudor (Mary Rose) and Etruscan shipwrecks: application of analytical organic chemistry in archaeology', Analyst Journal, 5 (1987)

Schmidt, H-P. and Wilson, K. 'The 55 Uses of Biochar', Biochar Journal (2014) www.biochar-journal.org/en/ct/2 (Accessed 31 August 2017)

Schmidt, H-P. 'The use of biochar as building material', Biochar Journal, Arbaz, Switzerland (2014) www.biochar-journal.org/en/ct/3 (Accessed 31 August 2017)

Schmidt, H-P., and Taylor, P. (2014): 'Kon-Tiki kilns – flame curtain pyrolysis for the democratization of biochar production,' Biochar Journal, Arbaz, Switzer¬land pp. 14–24 (2014) www.biochar-journal.org/en/ct/39

Shacklady, H. Ulverston, *An English Market Town Through History* (Handstand Press, 2016)

Shackley, S. 'The economic viability and prospects for biochar in Europe: shifting paradigms in uncertain times' in S. Shackley, G. Ruysschaert, K. Zwart, B. Glaser (eds) Biochar in European Soils and Agriculture: Science and Practice pp. 205–226 (Routledge, 2016)

Stern, S. T. 'Nanotechnology Safety Concerns Revisited', Toxicological Sciences 101 (1), pp. 4–21 (2008)

Singh, R., and Kumar, S. (eds) Green Technologies and Environmental Sustainability (Springer, 2017)

Steiner, C., McLaughlin, H., Andrew, H., et al. U.S.-focused Biochar report: Assessment of Biochar's Benefits for the United States of America, Report United States Biochar Initiative, Center for Energy and Environmental Study (2010) http://www.biochar-international.org/node/1816

Thomas, R., 'Chronology of gunpowder: Explosives and the Waltham Abbey Royal Gunpowder Mills' https://www.royalgunpowdermills.com/history-heritage/chronology-gunpowder/

Tong, Z., Pullammanappallil, P., and Teixeira A. A. *How Ethanol Is Made from Cellulosic Biomass* (University of Florida, 2015)

Uhlmann, J. Heinrich, P. *The Soul of Fire – How Charcoal Changed the World* (University Books 1987)

United Nations Development Programme, Review of Policies and Regulations on Charcoal and how United Nations Development Programme, 'Review of Policies and Regulations on Charcoal and how to Promote a Systems Approach to Sustainable Charcoal Production and use In Malawi' (May, 2012) https://jobs.undp.org/cj_view_job.cfm?cur_job_id=29854

Yu, X-Y., Ying, G-G., and Kookana, R. S. 'Reduced plant uptake of pesticides with biochar additions to soil' Chemosphere 76(5), pp. 665–671 (July, 2009)

Tan, Z., Lin, C. S. K., Ji, X., and Rainey, T. J. 'Returning biochar to fields: A review' Applied Soil Ecology, 16, pp. 1–11 (2017)

Useful Addresses

SUPPLIERS OF KILNS OR RETORTS

Exeter Charcoal Retort
Carbon Compost Co. Ltd
167 Mincinglake Road
Exeter
Devon
EX4 7DS
Tel: Robin on 01392 431454 or 07515 683908
Geoff on 01392 274699 or 07966681676
Email: admin@carboncompost.co.uk

Four Seasons Fuels
Coneyhurst
Billingshurst
West Sussex
RH14 9DG
Tel: 01403 783379
www.fourseasonsfuels.co.uk

Hookway retort
www.hookwayretort.co.uk

Oxford Charcoal Company
Range of charcoal products and retorts
Worton
Witney
OX29 4FL
Tel: 07890 967599
www.theoxfordcharcoalcompany.co.uk

Pressvess
www.pressvess.co.uk

Woodsmith
Set up by Maurice Pyle, this company supplies
a range of kilns
7–9 Claremont Road
Whitley Bay
Tyne and Wear
NE26 3TN
Tel: 0191 252 4064
www.woodsmithexperience.co.uk

CHARCOAL COMPANIES

London Log Company
Suppliers of wood and charcoal, including
wood-smoking charcoal
4 Abbey Orchard Street
London SW1P 2HT
Tel: 0303 123 4500 or 0208 3144592
Mobile: 0753 999 5725
www.thelondonlogcompany.blogspot.co.uk

Rivenwood Coppice
Iain Loasby
Blue House Yard
London
N22 7TB
Tel: 07812 367 480
Email: iain@rivenwoodcoppice.com
www.rivenwoodcoppice.com

Wildwood Charcoal and Coppice
Alan and Jo Waters
Tel: 01243 778106

Yorkshire Charcoal
David Hutchinson
Brompton Moor House
Sawdon
Scarborough
YO13 9EB
Tel: 01723 864957

SUPPLIERS OF BAGS AND PACKAGING

Selway Packaging
Suppliers of wide range of bags and sacks, spe-
cialists in charcoal bags for British Charcoal
Units 3 & 4C
Paddock Road Industrial Estate
Paddock Rd
Reading
Berkshire
RG4 5BY
Tel: 0118 946 2333
www.selway.co.uk

Classic Bags
Bespoke printed designed bags
5 Silver Business Park
Airfield Way
Christchurch
Dorset
BH23 3TA
Tel: 01202 488144
https://madebyclassic.co.uk/contact

EQUIPMENT SUPPLIERS

A.T. Sack Fillers
Supply a wide range of sack filling machinery
within the UK and internationally
12 Biggin Lane
Ramsey
Cambridgeshire
PE26 1NB
Tel: 01487 711114
www.atsackfillers.co.uk/

POTENTIAL CUSTOMERS

The British Orchid Growers Association
Orchid growers particularly favour horticultural biochar as part of an orchid-growing
medium; BOGA provides some useful links
to orchid growers and nurseries supplying
orchids
www.boga.org.uk/links.html

Water Gardeners International
Another source of useful links to organisations
and groups that use horticultural biochar
www.watergardenersinternational.org/clubs/
main.html

ORGANIZATIONS THAT SUPPORT CHARCOAL BURNING

Bill Hogarth Memorial Apprenticeship Trust
www.coppiceapprentice.org.uk
Email: info@coppiceapprentice.org.uk

British Biochar Foundation
www.britishbiocharfoundation.org

International Biochar Institute
www.biochar-international.org

National Coppice Federation
www.ncfed.org.uk

Small Woods Association
Green Wood Centre
Station Road
Coalbrookdale
Telford
TF8 7DR
Tel: 01952 432769
www.smallwoods.org.uk

CERTIFICATION

Forest Stewardship Council (FSC)
11-13 Great Oak Street
Llanidloes
SY18 6BU
Tel: 01686 413916
www.fsc-uk.org

PEFC – Programme for the Endorsement of
Forest Certification
PEFC UK Limited
Sheffield Technology Parks
Cooper Buildings
Arundel Street
Sheffield S1 2NS
United Kingdom
Tel: 0114 3072334
Email: info@pefc.co.uk

OTHER ORGANIZATIONS

Forestry Commission HQ
Public Enquiries
231 Corstorphine Road
Edinburgh
EH14 5NE
Tel: 0845 3673787
www.forestry.gov.uk

Grown in Britain
19 Common Road
Hanham
Bristol
BS15 3LL
Tel: 0117 958 2189
Email: enquiries@growninbritain.org

Health and Safety Executive (HSE)
Head Office
Health and Safety Executive
(1G) Redgrave Court
Merton Road
Bootle
Merseyside
L20 7HS
Tel: 0845 345 0055
www.hse.gov.uk
www.hse.gov.uk/reach/resources/
reachsds.pdf

Index

abrasion 12
absorption 12, 145
acetic acid 50, 79, 117
activated charcoal 11, 12, 82
activated charcoal 154
adder for luck 30
adsorption 12, 145
alchemy 8
alder 18, 78
alder buckthorn 18
alkaline 73
Allonby, Jack 22, 30, 38
Alsace, France 44
Amazon 20, 71, 143, 153
animal feed 155
apple 133
Armstrong, Lyn 23, 30
artist charcoal 155
ash 11, 50, 58, 73, 117, 148,
ash die-back 147

Backbarrow 17, 30, 78
Bagging 46, 100–17
bagging plant 112
bags, *see* packaging
barbeque, 114, 128–41, 132
barrel kiln 52–9
batch retorts 79
beech 133
binchotan 12, 160
biochar 12, 58, 71, 101, 115, 126–7,
    142–53
bio-engineering 153
biofuels 12, 142, 144
biofuelwatch 150
bioregional charcoal company 114
black powder 18
blast furnace 16
blue smoke 70
brands, *see* brown ends 48
brick kiln 20, 22, 75–7, 80
briquettes 128, 130, 137
British Biochar Foundation 122
British standards 123
bronze age 16
brown ends 48, 56, 107
building 157
burnishing 157

cahoon nut 84
carbon 9
carbon capture 150
carbon dioxide 142
Carbon Gold 95, 145
carbon sequestration 142, 143, 150

cation exchange capacity 147
cellulose 26
certification 124–6
changing chimneys 69
charboard 157
charcoal burners' hut 22
charcoal feedstock 26
Charfest 30, 40
chemical activation 155
chimney 35, 52, 67
chimney lighter 136
climate change 77
CO_2 8, 150
CO_2 cycle 151
coal wood 34
coke 17
colliers 17
compaction 147
comparitive table of methods 168
continuous burns 81
conversion rates 23, 24
cooking 18, 95, 128
cooling 46, 50, 52, 56, 70
co-operative working 114
coppicing 16, 28, 34, 48
cords 34
cosmetics 157
crab apple 30
Crawley, Brian 30
cylinders 78, 80

decaffeination 158
destructive distillation 12
diamond 10
distillates 11, 78–9
dog biscuits 158
double height kiln 52
Dow, Michael, film-maker 30
drum barbeque 134
dry distillation 12
dust masks 103

earth kiln or earth burn 20, 21, 22,
    30–48
eco firelighters 136
electromagnetic radiation 157
embalming 18
energy generation 86
enzymes 148
europaisher kolerverein (ekv), european
charcoal burners 19, 32, 49
European Biochar Foundation 143
Exeter retort 89 - 96
explosives 79

Fairtrade 126
felling 29
filtration 158
fines 144
fire brigade 119
firebox 86, 93, 97
fireworks 159
fish farming 159
flammability 12
flue, *see* chimney 35
food supplements 159
foodstuffs 159
free burn 55, 67
FSC 124

gasification 96
glazed paper 107
gold extraction 159
gourmet charcoal 114, 132, 160
grading 71, 105–7
graphite 10
green wood 107
ground reclamation 145
Grown in Britain 125
gunpowder 13, 17, 18, 78

hair shampoo 160
hand warmers 160
Hanna, Sam 30
hardwood 27
hazel 133
health and safety 118, 120, 122, 135
health warning 24
hearth 21, 32, 63
Hookway retort 25, 97
horizontal gas-air vortex 72
horizontal retorts 81
horticultural use 145
humidity regulation 157
humus 148
hygroscopic 11, 147

incense 162
ink 160
insecticide 18, 97
insulated 77
insulation 157
insurance 120
International Biochar Initiative 142
internet sales 114, 117
iron age 16
iron age bloomery furnace 14, 16
Ithaka Institute 143

kebabs 139

Kelley, Don 24, 33
Kenya 129
kindling 50, 54, 64, 71, 93
Kon-tiki kiln 20, 26, 52, 58, 71–5, 120

legislation 118–27
Lembach 19
lids 65
light bulb 158
lighter fluid 136
lighting 38, 54, 64, 93, 134
lighting fuel 114
lignin 26, 27
Lloyd, Walter 60
lumpwood 128, 130

Malawi 129
markets 101, 115, 117
mass catering 139
medical use 162
metal kiln 20, 23, 52–75
metallurgy 13, 16
methanol 18
Missouri kiln 77
moisture content 27, 47
motty peg 35
mould, *see* sieved soil 37
mycorrhizae 148

nanotechnology 162
naphtha 18, 48
National Association of Charcoal Burners 23

oak 133
off setting carbon 150
organic gardening 164
Otzi 16
Oxford retort 86
oxygen 11, 20, 24, 32, 46

packaging 101, 107, 117
par char, *see* brown ends 48
paraffin 64
Parr, Mark 132
paths 164
PEFC 125
pellets 164
pit kiln 20, 49 -50
pitsteads, *see* hearth 21, 30
planning permission 118
plastic bags 107
platform 63, 90
poisoning 162
porosity 147
ports 59, 61, 69, 70
prehistoric artwork 13
Pressvess standard retort 83

pyroligneous acid 18
pyrolysis 11
    flame curtain 20, 58, 73
    hot, flash 12, 150
    slow 12

rabbits 70
Rackham, Oliver 16
railway system 80
rain 50, 86
Ransome, Arthur 21, 22, 30
ratio of charcoal to timber 48
Reach legislation 121
refrigeration 164
re-ignition 46, 70, 105
research 145
restaurant grade charcoal 131
retorts 18, 20, 22, 24–5, 77
rhododendron 164
riddle, *see* grading 71
risk assessment 120, 127
rocket stove 97

sacrificial wood 46, 52
Saint Alexander of Comana 19, 30, 40
saltpetre 18, 78
sammel, *see* soil 37
sand, *see* soil 50, 70
scales 111
sealing 111
semi-conductors 164
shanklins 34, 35
shattered charcoal 27
shingling 37
silver birch 133
smelting 14, 16
smoke 48
smoked food 138, 141
softwood 27
soil 37, 50, 52, 56, 61
soil carbon 150
soil decontamination 157
soil improver 101, 117, 142, 143, 148
soil microbes 148
soil organic matter 153
soil pH 148
splitting logs 27, 62, 87
stacking 64
stapler 112
statutory nuisance 119
steam 46, 56
steam activation 155
steaming 138
sulphur 18, 78, 128
Superchar 100/500 96
Sussex Weald 16
sweet chestnut 133
syngas 96

tarpaulin 61
tattoos 164
teeth cleaning 164
terra preta 142
textiles 164
thermometer 93, 94
timber, logs 21, 26–9, 34, 49, 55
tools for earth burns 34, 35
torrified wood 12, 131
tractor 63, 64
trolleys 86
turf 37, 40, 43, 54, 58
twin chamber retort 85

UK biochar research centre 143
Ukrainian ck-2 retort 84
Ukwas 125
United States Biochar Initiative 143

vertical retorts 82
viper kiln 72
volatile oils 18, 24, 48, 67, 87
volume 108, 123

warm up exercises 121
water 24, 34, 46, 73, 99
water divining or dousing 33
water filters 166
Waters, Alan 30, 38
Watson, Richard 78, 80
Weald and Downland museum 30
weather 115
weathering 148
weighing 108, 123, 123
wheel rim barbeque 134
white charcoal, *see* binchotan 12
wild cherry 133
willow 18, 78
wind chimes 166
windbreak 38
wood cradles 84
wood gas 18, 24, 43, 67, 78
wood oils 18
wood pitch 18
wood processor 57 - 59, 62, 63
wood tar 18, 22
wood vinegar 18, 22, 164
woodland management 28
wool combing 17

xylem 166

yields 46, 166

Zimbabwe 167